Using Science and Technology Information Sources

by Ellis Mount and Beatrice Kovacs

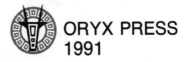

ORYX PRESS
1991

The rare Arabian Oryx is believed to have inspired the myth of the unicorn. This desert antelope became virtually extinct in the early 1960s. At that time several groups of international conservationists arranged to have 9 animals sent to the Phoenix Zoo to be the nucleus of a captive breeding herd. Today the Oryx population is nearly 800, and over 400 have been returned to reserves in the Middle East.

Copyright © 1991 by
The Oryx Press
4041 North Central at Indian School Road
Phoenix, AZ 85012-3397

Published simultaneously in Canada

Printed and Bound in the United States of America

∞ The paper used in this publication meets the minimum requirements of American National Standard for Information Science—Permanence of Paper for Printed Library Materials, ANSI Z39.48, 1984.

Library of Congress Cataloging-in-Publication Data

Mount, Ellis.
  Using science and technology information sources / by Ellis Mount and Beatrice Kovacs.
    p. cm
  Includes bibliographical references and index.
  ISBN 0-89774-593-0
  1. Science—Information services. 2. Technology—Information services. 3. Information storage and retrieval systems—Science. 4. Information storage and retrieval systems—Technology. 5. Data base searching. I. Kovacs, Beatrice. II. Title.
  Q224.M68   1991   90-21836
  507.2—dc20   CIP

Contents

Preface

There is little or no debate in society about the powerful influence of science and technology on our lives. The sources of information about these disciplines are themselves complex and thus worthy of careful study. Formal courses of instruction about sci-tech information sources have been a feature of the curricula of library schools since the 1930s. For many years science librarians, engineers, and scientists have relied on informal ways of learning about these sources, but these hit or miss attempts are inadequate for gaining a good understanding of the nature and use of the many sci-tech sources available.

Although society as a whole may recognize the importance of sci-tech information, there is a wide gap between the practicing scientist or engineer and the layperson in the understanding of these sources of information. Many students in library schools are apt to be at the level of the layperson when it comes to science and technology information sources because so few students in such programs have had much training in these subjects. As a result, library school courses concerned with the literature of science and technology must devote a considerable amount of time to teaching students about the nature of the many types or formats of sci-tech information that exist, such as the characteristics of patents, technical reports, or laboratory notebooks.

The purpose of this book is threefold: 1) to serve as a textbook in schools of library and information science for courses in science literature (as they are usually called), by presenting descriptions of the many sources of information on science and technology; 2) to aid librarians already involved with science and technology libraries, as well as users of sci-tech information, in gaining a better understanding of the characteristics of information sources that may be little known to them; and 3) to assist librarians with collection management of sci-tech literature. In view of the relatively high cost of sci-tech literature, this last point is most important. Guidelines for collection development, weeding of collections, and actual selections must be based on a good understanding of the many sci-tech information sources that exist.

This book is *not* a bibliography of sci-tech sources; several excellent books of that type already exist. However, most of those books devote little space to a description of the nature of the various sources that are cited. The citations provided by these bibliographies are helpful, but the average reader will not gain an understanding of the nature of the various formats for which the literature guides provide bibliographic citations. For example, the brief descriptions found in most guides to the literature may leave the reader wondering how a patent differs from a technical report or an annual review article. Similarly, the reader may not realize that some sources are much better choices for people needing a quick answer to a question than other informational sources, some of which, for example, may be better for more scholarly inquiries.

This book is not in competition with guides to sci-tech literature; it is rather a companion volume to such works. Library school courses could use this book as a text to accompany the guides to the literature that emphasize the listing of many sci-tech titles. Chapter 1 is a discussion of the general nature of sci-tech information sources; Chapter 2 is a description of the kinds of people who use such sources. The other 33 chapters describe each of the various formats in which these sources may be found (such as handbooks, patents, or technical reports). Where appropriate, chapters end with annotated citations of a few key examples of what the compilers feel are outstanding titles. No effort has been made, however, to duplicate the extensive number of citations found in traditional guides to sci-tech literature. In general, all examples are in English. Although printed publications predominate among the examples of outstanding titles selected, care has been taken to include some important nonprint sources of information, such as computerized databases, CD-ROMs, and other graphic forms. Each of these chapters summarizes in brief paragraphs the characteristics of each information source, such as "Significance," "Physical Characteristics," or "Availability." The remainder of each chapter is devoted to a fuller discussion of the nature of the information source.

Appendix I presents examples of typical reference questions, along with an analysis of the best type of sci-tech source to use for aiding the inquirer. Little has been published on this topic, a problem that faces each librarian dealing with users in all types of sci-tech libraries. Appendix II is a glossary that provides brief definitions of the most important sci-tech information source formats. A thorough index facilitates location of specific items.

We would like to acknowledge the assistance and encouragement given us by our respective classes in science literature at Columbia University, School of Library Service, and at the University of North Carolina at Greensboro, School of Education, Department of Library and Information Studies. In addition we thank Jean Martin for a careful review of the manuscript, providing us with many suggestions for improvement. We also appreciate the assistance given by Susan Klimley. Last, and certainly not least, we thank our respective families for their patience and support during the preparation of the book.

PART I

Introduction

Chapter 1
The Nature of Science and Technology Information Sources

Sources of information about science and technology are quite numerous, are sometimes very expensive, are found in a variety of formats, and are available in several levels of difficulty for different audiences. A good understanding of the nature of these information sources is indispensable not only to the scientist and the engineer but also to the student trying to master them and to the library professional who works with them. This chapter discusses the general characteristics of these materials, including both print and nonprint information sources.

FORMATS

Sci-tech information sources appear in a number of different formats, including patents, journals, handbooks, and technical reports. Some sci-tech formats are not found in other disciplines, such as patents, for example, which have little or no relationship to literature or the arts. On the other hand, poems, as a format, have little or no ties to sci-tech sources.

One traditional way of classifying information sources in any discipline, including science and technology, is to divide the various formats into two classes: primary and secondary.

Primary Sources

Primary sources can be defined as those that present or record certain information for the first time, as distinct from those sources that deal with information previously recorded or published. Examples of primary sources in science and technology include most monographs, patents, technical reports, certain types of journal articles, dissertations, conference papers, personal correspondence, and diaries, plus nonprint sources that include maps, audio or video recordings, blueprints, charts, and photographs.

One of the most important features of primary sources is the unique nature of the information they contain. In order to know exactly what a primary source has to offer on a subject, it is necessary to examine it directly, rather than rely on another person's interpretation of the information. Studying the original source would eliminate errors that might creep into efforts to present the information using a second-hand version. However, facsimiles of an original source may be necessary for economic or technical reasons. For example, the diary of a famous scientist might be literally unreadable or be on very fragile paper, unsuitable for general perusal. Thus reproductions of many primary sources must be used at times. One primary source may contain valuable information that might escape the recognition it deserves for years, while another such source may achieve more or less instant acclaim as soon as it is made known.

Secondary Sources

These are sources that compile, analyze, synthesize, or edit primary sources or other secondary sources. Examples include textbooks, encyclopedias, handbooks, reviews, guides to the literature, translations, abstracting and indexing services, and bibliographies.

The chief value of secondary sources is in reducing the workload of researchers who may have insufficient time or ability to glean important data from primary sources and who may want simplified versions of original data. Another attribute is that some secondary sources, such as indexing services, help searchers locate materials that were previously unknown to them. Even if they could take the time to study certain original sources, searchers might not be aware of what sources would be worth examining and they might not even be able to locate copies of the original sources to study. What secondary sources lack in original data is made up for in their usefulness to the average user.

It should be evident that both primary and secondary sources are necessary for efficient and effective research. Sometimes exact knowledge of an original bit of information is required, while in other instances the time-saving quality of secondary sources is far more important. This book will describe and provide examples of both types of sources.

LEVELS OF TREATMENT

Since society is made up of people whose knowledge of sci-tech subjects varies a great deal, it is fortunate that sci-tech information sources collectively cover a wide spectrum of knowledge, ranging from very simple approaches that a child could understand to arcane, high-level treatments that perhaps only a small number of scientists could clearly grasp. It is important for librarians to know how to select sources prepared for different types of audiences, be they knowledgeable adults, laypersons with little understanding of sci-tech subjects, or children with a limited command of language.

There is an abundance of basic materials written for children in the primary and secondary school grades, and there are also numerous nonprint sources such as films, slides, and recordings published just for their needs. The layperson does not seem to be as well served as children. A slow improvement has occurred in recent years, but the coverage of disciplines is spotty. A selection of sci-tech books written for the layperson, compiled by Ellis Mount and Barbara List, has shown that over the years certain topics have been very popular with book publishers, while other subjects are rarely treated.[1] For example, astronomy and astronautics are always well represented in the new books each year, but some other disciplines, such as chemistry or some of the engineering fields, get only slight attention by publishers. Publishers need to publish a wider range of sci-tech books for laypersons.

QUANTITY PRODUCED

The well-known annual survey of book quantities and prices done by *Publishers Weekly* shows that nearly 11,000 hardcover and trade paper books relating to science and technology were published in the United States in 1988; most of them (83%) were new titles and the rest new editions.[2] This is not an inordinate number when compared with the grand total of 52,000 books in all disciplines (both new titles and new editions) published that year in the U.S. The breakdown by subject was as follows:

Sci-Tech Book Output 1988

Subject	New Titles	New Editions	Total
Agriculture	561	102	663
Medicine	3,187	675	3,862
Science	3,091	643	3,734
Technology	2,209	457	2,666
Totals	9,048	1,877	10,925

As for periodicals or journals, the quantity of titles currently being published varies a great deal; each year sees the inauguration and demise of many titles. Close to 100,000 sci-tech journals have existed at one time or another. The listing of current sci-tech journals in the 1988 issue of *Scientific and Technical Books and Serials in Print,* which does not claim to be exhaustive, consists of some 17,000 titles selected on an international basis.

In recent years the quantity of publicly available online databases dealing with science and technology has grown to number in the hundreds. Some cover all aspects of these disciplines, while others are less broad in their scope.

COSTS

This is another characteristic having wide implications. Sci-tech books for children currently have a low average of $13, whereas, when the titles listed by Mount and List were totaled informally, they showed the average price of sci-tech books for laypersons published in 1988 to be $24.41. Sci-tech books aimed at a lay audience tend to be much cheaper than those written for professionals. The annual *Publishers Weekly* survey previously mentioned shows a considerable increase in the cost of hardcover sci-tech books (those not written especially for the layperson) in the past 10 years, averaging around 155% from 1977 to 1987. On the other hand, hardcover books in all disciplines (such as art, literature, fiction, and law) had an overall increase of only 17% in that period. By contrast, some of the sci-tech categories increased from 7% to 15% in one year alone (1987–88), as shown below:

Sci-Tech Book Prices

Subject	1977 Prices	1987 Prices	1988 Prices
Agriculture	$16.24	$46.24	$49.36
Medicine	$24.00	$57.68	$66.59
Science	$24.88	$62.16	$66.91
Technology	$23.61	$60.24	$65.26

The figures for 1989 show increases in the range of 2.4% to 8.4% for these four categories.[3]

Subscriptions to scientific journals are also costly; some cost more than $2,000 per year. An annual subscription to the highly esteemed *Chemical Abstracts* costs more than $11,000.

The cost of online searching of sci-tech databases, however, varies greatly with sponsorship. Those owned by government agencies are offered at much lower prices than most of the others. Charges as high as $200 per hour are not unusual for some privately owned databases.

TIMELINESS OF DATA

In some disciplines scientists and engineers need to have access to the most recent information on a given topic. More accurate information supersedes older data and newly developed methods replace outmoded ones. Current information can be critical. Therefore, the value of certain information formats that allow announcement of new data is great. Such formats include serials, technical reports, and online databases.

Of the formats that contain current information, serials occupy an extremely important place in sci-tech libraries. They come in several types and are known by a variety of names. Current scientific information for the public is usually found in general interest serials commonly known as *magazines,* such as *Today's Health.* Somewhat more technical presentations and articles, designed for readers with some background in the disciplines involved, appear in what are generally known as *periodicals,* as exemplified by *Scientific American.* The most advanced level of information on science and technology appearing in serial format is found in what are called *journals.* Newspapers and newsletters are more limited in scope but appeal to a number of levels of understanding.

Less frequent serial publications, which still provide somewhat current information, are annuals and yearbooks as well as those sources known as "pseudo-serials." The latter are actually monographs (separate publications) issued in new editions on an annual basis, usually handled as serials in libraries. These less frequent publications should not be confused with monographic series, which are individual books issued on an irregular basis and tied to each other simply by virtue of sharing a common subject area and series name. More detailed descriptions of serials may be found in some of the chapters that follow.

Some relatively old information may still be valuable in certain disciplines. Not every experiment or finding is superseded by newer work. In some disciplines, such as geology or mathematics, it is not uncommon for work done in previous centuries to remain valid today. A geologist in the 1800s may have written a description of some rock formation in a remote area of the earth that no one else subsequently studied. Thus, the original description of the area might be the only known record on that topic. Likewise, certain mathematical studies written decades ago might still be perfectly acceptable to present-day mathematicians.

UNIVERSALITY

One of the outstanding characteristics of sci-tech data is its imperviousness to political boundaries, its universality. A chemist in Finland uses the same laws governing chemical reactions as a chemist in Philadelphia. An experiment to measure the strength of a bar of steel in Detroit would give the same results if done (with the same materials and same techniques) in Moscow. In some ways scientists and engineers have more in common with their counterparts in foreign countries than is the case with most other professions.

Although the basic principles and techniques of science and technology are the same in laboratories and factories around the world, it is obvious that some countries are more advanced scientifically and technically than others. Many areas of the world, particularly Third World countries, are woefully lacking in modern equipment because of the high cost of that equipment. The body of sci-tech information is, however, being updated daily by the contributions of sci-tech personnel on a global basis, including those from Third World countries. This universality of science and technology is one reason it is so important for scientists and engineers to keep abreast of developments in other countries. Unfortunately, in actual practice, only a small percentage of sci-tech researchers and practitioners come close to being fully informed about foreign research in their fields.

CUMULATIVE NATURE

One feature of science and technology that is lacking in many other disciplines is that present-day sci-tech knowledge is built upon information gained in previous years and centuries. No scientist or engineer needs to reinvent basic information readily available in the various information sources. "Reinventing the wheel" has become a standard phrase for indicating that some project or effort is worthless because it has already been accomplished. It is appropriate that the phrase is couched in technical terms, since there is so little excuse for not knowing major developments of the past. If, for some reason, earlier developments could not be used to aid present-day efforts, our progress in science and technology would be slow.

There have been, however, many instances in which important developments were overlooked or not accepted for many years due to a variety of causes. One reason was personal prejudice against particular people, causing others to ignore the originators' papers and publications. A more common reason was that the original public disclosure of the information was in a publication that was not widely read; consequently, the information was hidden until rediscovered, sometimes years later. Still another reason was that some discoveries were so contrary to contemporary thinking that scientists and engineers were unwilling to accept them.

A classic example of the latter case is the paper written in 1869 by the Russian chemist Dmitri Mendeleev in which he showed how the chemical elements could be arranged by their atomic weights to form a table, with elements with similar characteristics falling into the same columns. Mendeleev even predicted the discovery of elements that would fill the gaps in his table. His paper was generally ignored or widely criticized by other chemists. It was not until some four years after he predicted the existence of certain elements that one such element was in fact found to exist, followed by still others in the 15-year period following publication of his original paper. Only then was Mendeleev's concept of the periodic table beginning to achieve general acceptance; in later years he received worldwide acclaim (except in his native Russia, which objected to his liberal political views).[4]

INFORMAL MEANS OF COMMUNICATION

Many studies of the ways in which scientists and engineers obtain information have confirmed the great use they make of informal methods of communication, particularly oral communication. This is not surprising, since most people follow the same pattern.

There are many advantages to oral communication (often called the invisible college), namely ease and speed of communicating, the possibility of obtaining current information not yet published, and the possibility of learning bits of information that may never be published, such as details of certain events that were not suitable for publication. Another advantage is that oral communication is interactive, so that the questioner can ask for clarification or elaboration of certain points, an advantage that is often of great importance.

On the other hand, there are several disadvantages to reliance on oral communication. One is that most of us are not well enough acquainted with all the people who could best answer various queries. Granted that certain persons would be excellent sources, how many of us would feel comfortable calling them for information? Another problem is accuracy. An oral transmission of many numbers and/or formulas would be quite likely to contain a number of inaccuracies. In some cases use of a graphic approach would be the best way for a person to understand a particular bit of information, but informal transmission of pictures or graphs is still not a common part of our telecommunication systems. Finally, even experts might recall bits of information incorrectly and thus unwittingly pass on errors when called upon for aid.

Informal communication includes personal correspondence, which can be quite valuable. Biographies of famous scientists and engineers often reveal the voluminous correspondence they carried on with their colleagues. Again, acquaintance with the person providing the answers sought is generally required, not an easy matter for those lacking a prominent standing in a field. On the plus side, tabular and graphic data can be transmitted somewhat easily through correspondence, and the process can be relatively speedy, albeit slower than oral methods. Interaction is possible but is not as easily done as in oral methods. Telefacsimile (fax) is another means of quick transmission of textual and graphic data.

The rapid growth of computers has led to the development of still another type of informal communication, namely electronic mail (sometimes called E-mail). This involves networks that tie together operators of computers who have access to those systems. By virtue of this access, network members can send messages from their computers to the computers of other members. Speed of communications is much faster than mail and less expensive than using telephones, particularly for international communication. It is very common in certain sophisticated organizations for originators of internal documents to announce availability of their reports via E-mail, often eliciting several requests for copies of the documents. Another use is to put an inquiry on an electronic bulletin board, which often involves thousands of people, asking for guidance or help on a particular problem. For example, a request such as "Does anyone know who is working on algorithms for parallel memory allocation?" might result in a number of responses from knowledgeable people.

This chapter has aimed at giving the reader a better understanding of the rather complex nature of sci-tech information sources. The next chapter deals with the individuals who use scientific and technical information and the remainder of the book will be devoted to a description of particular formats in which this information can be found, along with a selection of representative examples of each format.

REFERENCES

1. Mount, Ellis; List, Barbara. Best sci-tech books of 1988. *Library Journal.* 114(4): 39–46; 1989 Mar. 1.
   Presents a selection of what the compilers judged to be the best 100 books (published in 1988) for sci-tech collections in general libraries. Most are written for the layperson.

2. Grannis, Chandler. Titles and prices, 1988; final figures. *Publishers Weekly.* 236(13): 24–27; 1989 Sept. 29.
   A highly esteemed annual survey of the costs and quantities of American books published in different subjects or disciplines.

3. Grannis, Chandler. Book price averages 1989. *Publishers Weekly.* 237(40): 16–18; 1990 Oct. 5.
   Increases from 1988 to 1989 were relatively moderate.

4. Marks, John. *Science and the making of the modern world.* London: Heinemann; 1983: p. 311–313.

A readable book on the history of science, one chapter is devoted to the accomplishments of nineteenth-century scientists, including Dmitri Mendeleev's periodic table of the elements.

Chapter 2
The People Who Use Scientific and Technical Information

Just as it is important to understand the nature of informational materials in science and technology, it is important to understand the background of the people for whom these materials were produced. The terms "science" and "technology" refer to such a wide variety of subjects, that it is almost impossible to create a comprehensive list. The subjects include basic information about some aspect of nature and also the most technical and advanced application of a hypothetical theorem. To understand the materials, one needs an awareness of their audience.

Many of the informational materials produced in science and technology are geared toward individuals who have attained a particular level of education or experience. This is a relatively easy concept to accept, but a difficult concept to define. People who use sci-tech materials must be aware of the expectations of background and knowledge base that the creators of that material have for the audience. Often, the expectations are identified somewhere in the beginning of the book or other material, often in an introduction or beginning chapter. In most cases, the format in which material is presented can be an indicator of the degree of background and knowledge the reader is expected to bring to the material.

Many sci-tech materials are produced for those with little or no scientific or technical background. These materials can be found in bookstores, supermarkets, and public libraries. Very often, such materials are extensively illustrated and are usually written with a minimum of scientific jargon. Many of these materials are designed to be used by children, to introduce some aspect of science or technology. Others are designed to be used by the curious adult who might want to begin a hobby or satisfy a simple need for information. These sources usually contain simplified descriptions of the important aspects of the subject, with glossaries of terms for the uninformed reader, and perhaps lists of materials or organizations for further information.

Materials are also created in sci-tech disciplines to support educational programs. These are tools for teaching the major aspects of the subject in structured situations; textbooks are prime examples of this type of material. Textbooks are written for pupils in elementary schools as well as for students at universities. Obviously the language of the text and the information covered will vary with the target audience for the work. Audio and visual materials are often created for these types of audiences because a good deal of information can be shared with a large number of individuals. It is important, when seeking to acquire such materials, that the target audience be identified to see if the titles will meet a current need.

The next level of background and knowledge to consider is the post-baccalaureate researcher (in terms of academic disciplines) or the post-apprenticeship professional (in terms of many technical disciplines). Depending on the science or technology, works will be written with these specific individuals in mind. The type of language or the lack of explanatory data clearly marks these materials as inappropriate for the novice or beginner.

The graduate or post-doctoral researcher who is involved in very specific, very technical, or very esoteric work in an area of science is not concerned with explanatory material. He or she wants specific pieces of information as soon as possible to solve some problem or answer some question. The technological counterpart to the post-doctoral researcher is the experienced professional who is involved in the design of advanced systems or procedures. This person, too, is generally not interested in explanatory materials, but rather in facts or specific information to aid in the resolution of a problem or to answer a question. However, highly skilled researchers are occasionally asked to work outside their normal areas of expertise; such people invariably ask for help in locating a few key books or reference works in the new field to help them get acquainted with basic studies before attempting to go further in the field.

Consideration should also be given to persons who are employed in some aspect of scientific or technical work without an extensive educational background. These people often become laboratory assistants or technicians and need information sources that will aid them in the performance of their duties. Materials necessary for their jobs may be less technical than for the researcher or the scientist, yet may contain detailed job-specific information that would not be helpful to the general public.

Such generalizations can be made regarding the probable audience for many of the material formats in sci-tech literature. For example, certain types of information will most often be of use to the expert, such as technical reports, dissertations and theses, patents, handbooks, standards and specifications, laboratory notebooks, and preprints. For the advanced student of a subject, the most useful materials will include conference papers, monographs, histories, translations, annuals and yearbooks, and reviews of the literature. The person who is new to a sci-tech discipline will need dictionaries, encyclopedias, thesauri, guides to the literature, textbooks, histories, bibliographies, and audio-visual materials. Technicians and laboratory assistants will need textbooks, manuals, handbooks, thesauri, standards and specifications, and audio-visual materials.

All of these groups will also find periodicals, newspapers and newsletters, government publications, and computerized information helpful. These lists are by no means exclusive; the expert may very well have a great deal of use for a dictionary, and the novice might find standards and specifications to be very helpful. The preceding list is meant only to give an indication of the usual expectations for use of such materials.

There are many individuals who can be classified in more than one of these categories. For example, there is the post-doctoral researcher in chemistry who decides to develop a hobby, such as geology. There is the archaeologist who wants to learn about underwater welding, and the ornithologist who becomes interested in astronomical observations. There are materials designed for most of these individuals and situations.

PART II

Primary Sources of Information

Chapter 3
Conference Literature

HIGHLIGHTS OF CONFERENCE LITERATURE

Description: Conference literature consists of papers read at meetings or conferences, as well as the collection of such papers into a single published volume.

Significance: Papers prepared for presentation at conferences are similar in several ways to journal articles--they are relatively current, are usually specific in scope, and can be the first public disclosure of certain information. Many important sci-tech developments are first made known at such events.

Quantity: There are thousands of conferences held each year in the United States, with thousands more held in other countries. At each meeting, dozens of papers are usually presented. Gloria Moline has estimated that more than 150,000 papers related to engineering alone are published every year.[1]

Physical Characteristics: Most conference papers are similar in appearance and length to periodical articles. In bound form they have the appearance of a bound periodical volume.

Availability: Many papers can be obtained by writing the author for a copy (not always a successful effort). The papers may be made available at the time of the meeting (though usually only in abstract form), or in a compilation prepared for sale some time after the meeting, or they may appear later as separate periodical articles.

Retrieval: There are a few special indexing services which are geared solely to identifying conference proceedings and individual papers of recent vintage; they have a broad scope in terms of the number of meetings covered. Papers from certain important conferences are indexed regularly by major indexing and abstracting services (or their online equivalents). Such services are quite selective in their coverage of such material, being devoted chiefly to journal articles. Papers from other conferences may go unnoticed if they are not indexed by any of these services. Some conference papers are published in bound volumes which collect all the papers given at a meeting; publication increases the likelihood that such papers will be indexed.

Intended Audience: Conference papers generally appeal to those who are well grounded in the subject matter. Like most periodical articles, conference papers are not written on the tutorial level for neophytes.

Scope: Like journal articles, conference papers can cover any of a variety of knowledge levels for a particular subject. Most papers deal with details of some specific topic.

NATURE OF CONFERENCE LITERATURE

On any one day in the year many engineers and scientists are attending conferences, symposia, workshops, panel discussions, conventions, colloquia, congresses, and seminars.

At these meetings, hundreds of papers written by specialists are read. These meetings may be on a local, regional, national, or international level. Sponsors might be a professional association, a commercial firm, a government agency, an educational institution, or perhaps a combination of various groups. The collections of papers that come out of these meetings are themselves known by a variety of names, including proceedings, transactions, abstracts of papers, and papers presented. Probably "conference proceedings" is the most common way of referring to collections of papers presented at a formal meeting. Variety—whether in names of conferences, sponsorship, or techniques for dissemination of information presented—is a hallmark of sci-tech conferences.

Many are routine reports of research, not of earthshaking importance. On the other hand, some of the most significant developments in a field are first announced to the world at conferences, and important plans for the future are often made public there. For example, the first success in splitting the uranium atom was announced at a conference in the 1930s. A few meetings provide such valuable information that the best papers are reported on the front pages of major newspapers. Obviously not every paper or every panel discussion has great significance to the general public; nevertheless they are apt to be very useful to the sci-tech world.

Keeping abreast of conference papers and discussions is not an easy task for several reasons—there are so many meetings that not all of the papers and discussions are available in printed form, and even those in printed form are not easy to obtain. The availability of papers varies greatly from conference to conference. At one extreme is the conference at which the papers are not collected and printed, and the authors are forced to seek publication of their papers in periodicals, often long after the original presentation.

Somewhat better is the practice of distributing copies of papers to attendees. Some meetings are so well organized that printed sets of abstracts of all scheduled papers are given to attendees at the conference registration desk. These sets are frequently available for later sale to people who did not attend. Another variation is the practice of selling audio-cassette versions of papers. Panel discussions, a frequent feature of conferences, are particularly well suited to being recorded on cassettes, since such programs often involve an informal give-and-take between the panel members and the audience, unlike a formal paper prepared in advance. If a meeting were important enough, it is possible that major speeches and panel discussions might be recorded on videotapes.

Some organizations that sponsor meetings publish selected papers in their journals; others make a practice of publishing collections of either partial or full versions of the papers. However, months or even years often go by before the printed papers are published. Sometimes they are issued in hardbound books, which might be part of a long series devoted to that particular conference.

For many years certain indexing and abstracting services have indexed selected conference papers, and some indexes are devoted entirely to coverage of printed conference proceedings. One is the *Directory of Published Proceedings,* which has a separate section devoted to printed collections of papers from sci-tech meetings. In recent years another way of learning about papers given at meetings has been through certain online databases that index sci-tech papers in various disciplines; an example is CONFERENCE PAPERS INDEX, which has a file of more than one million sci-tech conference papers.

Usually the indexes provide only a citation and brief abstract, but at least the searcher is alerted to the existence of papers of interest. An example of a printed indexing service of this sort is *Index to Scientific & Technical Proceedings & Books.* It allows users to locate conference papers by titles, authors, or key words. One firm has been selling copies of papers as its main function, but complete coverage of all meetings is apparently not feasible, an understandable situation in view of the large quantity of papers given annually.

There are several ways of becoming aware of future meetings that may be of interest to scientists and engineers. For example, a few serial publications consist entirely of lists of such meetings. Some of these journals restrict their lists to events in the United States, but at least one is international in scope. Many sci-tech periodicals include, along with their traditional articles, a brief calendar of future meetings related to their particular disciplines.

The variety of ways to obtain data presented at conferences can create many difficulties in information access, but the value of that information makes it important for researchers to understand how to overcome those difficulties.

TYPICAL EXAMPLES OF SOURCES OF CONFERENCE LITERATURE

CONFERENCE PAPERS INDEX. Bethesda, MD: Cambridge Scientific Abstracts; 1973– . Updated bimonthly.
An online database available on DIALOG, it provides access to more than 100,000 papers presented at over 1,000 meetings each year. There are over 1,300,000 papers in the file. Besides being searchable by authors, titles, or names of conferences, the records show the availability of any publications that are issued in connection with the papers. Also available in printed format (with annual cumulation).

Directory of Published Proceedings. Series SEMT. Harrison, NY: InterDok Corporation; 1964– . 10 issues/yr.
Published in several series, the SEMT edition deals only with conferences devoted to science, engineering, medicine, and technology. Each proceedings volume is cited, with entries arranged by broad categories. There are indexes by editors, titles of conferences, and specific subjects.

Index to Scientific & Technical Proceedings & Books. Philadelphia: Institute for Scientific Information; 1978– . Monthly with annual cumulations.
Lists the titles of selected sci-tech conference proceedings as well as the titles of individual papers, including those published in journals as well as books. Has indexes by names of authors as well as key words from titles.

REFERENCE

1. Moline, Gloria. Secondary publisher coverage of engineering conference papers: viewpoint of Engineering Information, Inc. *Science & Technology Libraries.* 9(2): 47–61; 1988 Winter.
Describes the handling of conference literature at this outstanding abstracting organization. The issue also contains papers devoted to several aspects of conference literature, such as commercial publishing of conference proceedings, handling of this material at the Engineering Societies Library, retrieval techniques, and cataloging of conference proceedings.

Chapter 4
Dissertations and Theses

HIGHLIGHTS OF DISSERTATIONS

Description: A dissertation is a document presenting original research on a topic, written by a candidate for a doctoral degree. It has been carefully reviewed before acceptance by a committee of experts at a university and is generally written on a scholarly level.

Significance: Dissertations offer a source of carefully researched data, usually the product of several years of work. In most universities there are many more doctoral dissertations written in the pure sciences than in technology. Dissertations are not as well-known as other formats of sci-tech literature, such as monographs and periodicals.

Quantity: Currently there are approximately 35,000 doctoral dissertations written annually in the United States. Some 900,000 doctoral degrees have been granted in this country since 1861, the year the first one was awarded.

Physical Characteristics: Most dissertations are available either in print format or on microfilm.

Availability: The universities at which dissertations are accepted invariably have full-size copies in their libraries, but availability varies greatly from one school to another. Some schools make full-size copies available at the school or for interlibrary loans. Still others do not permit interlibrary loans and make their dissertations available only to purchasers of microfilm copies. Many schools will allow either loan or sale of microfilm copies, depending upon the borrower's request. A good source for a commercial copy (either in full size or on microfilm) is University Microfilms, Inc., which has an extensive file of dissertations and issues an index. In some instances, UMI requires purchasers to obtain a letter from the author that authorizes the making of a copy.

Retrieval: Dissertations are usually cataloged in libraries at the universities where they were accepted as well as in numerous indexing and abstracting services. Many are listed in OCLC and other bibliographic utilities. At least two online database vendors offer a file on dissertations, namely DIALOG and BRS.

Intended Audience: Dissertations generally are best suited for those involved in extensive research in a discipline. They are not written for the neophyte. They are much more important in some disciplines—geology, for example—than others.

Scope: Dissertations are usually narrow in scope, to permit a thorough analysis of the subject. A typical dissertation might present research on the nature and operation of an existing device or process, or it might disclose an entirely new theory about a process. A dissertation on a topic as broad as, say, bridges, or farm animals, would be almost unheard of. It is more likely that a dissertation would investigate the strength of bridges made with a particular arrangement of their structural members or the effect of a particular diet on sheep. If a person is interested in its topic, much can be gained by studying a dissertation.

NATURE OF DISSERTATIONS

One of the most valuable but at the same time most neglected sources of information about scholarly research and investigation is the doctoral dissertation, a carefully screened and well-supervised form of publication. It is more common in the pure sciences than in technology and certain forms of applied science, but the dissertation has an overall position of importance. It is meticulously scrutinized by a committee of faculty members (and sometimes outside experts) as to its accuracy and freedom from personal bias. It has to be unique, rather than retracing old research; has to be accepted by a special committee; and has to provide a well-documented account of its findings. Yet many scientists and engineers fail to make extensive use of such works when seeking information. The reasons for this situation are discussed below.

The first doctoral degree in the United States was granted in 1861 by Yale University; since that time, hundreds of thousands have been awarded in the U.S., amounting to nearly half a million by the 1970s. A decade ago nearly half the doctoral degrees in the United States and Canada were in the physical sciences, earth sciences, and life sciences; in the 1976–77 school year more than 15,000 out of some 32,000 doctoral dissertations granted were in these three disciplines.[1] Recent studies have shown that these proportions are still about the same. For 1986–88, dissertations in the sciences rose from approximately 44% to around 51% of the total (about 35,000 per year).[2]

One reason dissertations are little used by readers seeking information is the delay in the appearance of citations to recent doctoral dissertations in the regularly published literature. Only a handful of sources announce the completion of dissertations, and these sources are not noted for providing speedy, widespread notification to potential readers. Furthermore, dissertations in general must usually be purchased from a single commercial source, unless one wishes to pursue the laborious process of obtaining them from the originating school. Many dissertations will, in time (perhaps two years or so after their acceptance), appear as published books, but in many cases the age of the data can be a real handicap to sales.

The one good announcement source for American and Canadian dissertations is *Dissertation Abstracts International*. Foreign titles received little public notice in this country until 1988, when this publication began to include titles from Great Britain and the continent. This index is also available online. A comprehensive cumulation of dissertations cited by *Dissertation Abstracts International* up to around 1973 was issued that year, entitled *Comprehensive Dissertation Index*. A few abstracting and indexing services (some of which are also available as online databases) include selective coverage of dissertations in their respective fields of interest.

HIGHLIGHTS OF THESES

Description: Most theses are research papers required for certain master's degrees. They are not as common now as they once were. Generally they represent work done during a one- or two-year graduate program.

Significance: Theses are not as valuable as dissertations because they do not involve as much research and preparation. Nevertheless, they should be considered useful sources for certain purposes. As is the case with dissertations, theses are much more important in some fields than in others.

Quantity: Many schools no longer require master's theses. It has been estimated that approximately 2,500 theses are prepared each year in all subject areas.

Physical Characteristics: Most theses are available only as full-size copies of the original. Some sources may be able to provide them on microfilm.

Availability: In most cases the only source is the college or university where the paper was prepared.

Retrieval: Just about the only reliable index to master's theses is *Masters Theses in the Pure and Applied Sciences Accepted by Colleges and Universities.*

Intended Audience: Most theses are of interest chiefly to graduate students pursuing similar fields of study. In general, a thesis is not apt to command the same degree of interest as a dissertation on the same topic.

Scope: A thesis would probably cover less ground than a dissertation and might consist of a more general study of a given topic.

NATURE OF THESES

A master's thesis is a less-supervised and shorter piece of work than a dissertation, and the requirements for scholarly treatment of a subject are not as rigorous, nor does the thesis provide as much scope and depth of coverage. A thesis is generally a product of a one-year or two-year graduate program. This may reduce the value of theses to the scientist or engineer. Although many schools have eliminated the requirement for a thesis in their master's programs, some schools still make them mandatory.

Master's Theses is about the only source for locating theses; it provides bibliographic information and a subject approach, a rather rudimentary level of coverage. Most of the traditional indexing and abstracting services do not cover theses, another indication of the smaller amount of attention they draw. There is no central depository for them. They do not constitute a major source of information on science and technology, but they should not be overlooked.

TYPICAL EXAMPLES OF SOURCES OF DISSERTATIONS AND THESES

Comprehensive dissertation index. Ann Arbor, MI: University Microfilms; 1973. 37 vols.
    Provides information on more than 400,000 dissertations, arranged by broad subjects, then alphabetically by key words. Has indexes by authors and key words. There are separate volumes for each major discipline

Dissertation Abstracts International. Part B. The Sciences and Engineering. Ann Arbor, MI: University Microfilms; 1938– . Monthly.
    Previously limited to dissertations accepted in U.S. and Canadian schools, a growing number of European dissertations are included now. Arranged by broad subject (such as "Chemistry" or "Geology"). There are indexes by titles, key words, and authors. Available online as DISSERTATION ABSTRACTS, as well as on a CD-ROM disk, with both formats providing coverage from 1861 to date. Also includes citations to master's theses, based on theses listed in *Masters Abstracts* since 1962. An excellent source.

Master's Theses in the Pure and Applied Sciences Accepted by Colleges and Universities of the United States and Canada. New York: Plenum; 1955– . Annual.

An annual listing of master's theses prepared at more than 200 U.S. universities and colleges. Entries are arranged by broad subjects, such as "electrical engineering" and "chemistry."

REFERENCES

1. Subramanyam, Krishna. *Scientific and technical information resources.* New York: Dekker; 1981: p. 77–80.

Provides a listing of reference tools pertaining not only to domestic dissertations and theses but also to those prepared at foreign schools. A detailed account.

2. Personal communication from Patricia A. Kosobud, editor of the CDI Database at University Microfilms International, Ann Arbor, MI.

Chapter 5
Journals and Periodicals

HIGHLIGHTS OF JOURNALS AND PERIODICALS

Description: Journals and periodicals are familiar to everyone as publications issued on a regular basis and expected to continue indefinitely. Journals tend to present in-depth accounts of developments in the field(s) of interest to their readers, while periodicals tend to be less technically oriented.

Significance: Journals and periodicals represent one of the most important sources of information on science and technology because their data can be quite current, because they can be very specific in their treatment of subjects, and because there are so many of them in existence that a journal can be found on practically any subject in the sci-tech world.

Quantity: It has been estimated that at least 30,000 journals and periodicals devoted to science and technology are currently published.

Physical Characteristics: Journals and periodicals vary greatly in size, number of pages per issue, number of issues per year, and in other particulars. Just as varied is their quality of paper, attractiveness of type font, use of color, layout of contents, and the like.

Availability: Although there are a few exceptions, both periodicals and journals are obtained by paid subscriptions, placed either with subscription agencies or with the publishers themselves. There are few barriers to obtaining copies, although the rapidly rising cost of sci-tech serials has become a matter of great concern in recent years.

Retrieval: Because of the great number of periodicals currently being published, the only practical way to locate particular articles in journals and periodicals is through the use of periodical indexing and abstracting services or their online counterparts. The latter allow for searching by a variety of means, including authors, titles, key words, or assigned subject headings.

Intended Audience: Because of the great variety of serials being published, it is safe to say that there is something for everyone as far as periodicals and journals are concerned. Journals for practicing scientists and engineers are the most numerous, but a few periodicals are aimed at beginners or those with little or no background in these subjects. Some are designed for technicians or laboratory assistants. Some sci-tech journals have a potential audience of a few thousand readers; others might appeal to tens of thousands of readers.

Scope: Some journals and periodicals attempt to cover all areas of science and technology, presenting the highlights of new developments. More commonly, titles concentrate on a much more specific field of interest, such as welding, or organic chemistry, or quality control in manufacturing processes. The narrower the scope, the more detailed the articles can be.

NATURE OF JOURNALS AND PERIODICALS

If sci-tech librarians had to limit their collections to a single format, most of them would undoubtedly pick journals or periodicals as their choice. There are several good reasons why such publications are so important in sci-tech libraries:

- Journals or periodicals are able to publish new information more quickly than books. Some periodicals print new information only a few days after it is discovered; by contrast, most books require many months (sometimes over a year) between the completion of a manuscript and the appearance of the book in the market.
- Many titles are published, so it is possible to find one on almost any scientific or technical topic; quite often there are so many available that one must choose from several covering the same field.
- Journal and periodical articles can be very specific in their treatment of a subject, much more so than books, which tend to be broader in scope; books that are too specific often have very narrow markets and would be unprofitable for publishers to produce.
- Many abstracting and indexing services (including online databases) index and annotate periodical and journal articles; not every title is covered by an indexing service, but the better ones are often indexed by more than one service.
- Journals and periodicals are published with a variety of audiences in mind, ranging from children to top-level sci-tech researchers; there are few, if any, readers whose needs are not covered by some current title.

Sci-tech journals and periodicals have existed since 1665, when the French periodical *Journal des Scavans* began publication; it is still in existence, although in 1816 it changed its name to *Journal des Savants.*[1] Another important journal, which began later in the same year, is also still being published; the early volumes of the *Philosophical Transactions* of the Royal Society of London have been reprinted and are thus readily available. Since 1665, thousands of sci-tech periodicals and journals have come and gone. Derek J. De Solla Price prepared a graph showing that in 1750 there were only 10 sci-tech journals in existence in the world; this number increased to 1,000 by 1850 and to nearly 80,000 by 1950. By now the total number of sci-tech titles that have existed at one time or another is probably around 100,000.[2]

Types of Journals and Periodicals

There are several ways of classifying serials by means of the audiences for whom they are prepared. Some periodicals are aimed at the lay market. This audience prefers well-illustrated and clearly written articles. A prime example of this is *Scientific American,* which for many years has appealed to the average well-educated person; a measure of its popularity is its availability at newsstands in some cities. It is an ideal serial in which to find very clear discussions of sci-tech topics. Several other periodicals aimed at the same audience have appeared, but most of them are short-lived. For the person wanting short how-to-do-it articles, usually on technical rather than scientific topics, there are periodicals like *Popular Mechanics.*

Going up the scale of difficulty, the next class of periodicals is trade journals, devoted to such topics as welding, materials handling, design of products, and building trades. These are written for technicians, construction supervisors, and people in similar positions. Their articles are very detailed and technical and play a large role in educating their readers in new techniques and new products.

At the top level, as far as rigor of treatment goes, are professional journals, written for engineers and scientists who are interested in research as well as theoretical studies. A high percentage of these journals are published by professional organizations rather than by commercial publishers, although many fine journals in this category are commercially sponsored. Those published by professional organizations rarely have advertisements, often have good book review sections, and require submitted articles to be evaluated and recommended by selected, anonymous reviewers before being accepted for publication. Some societies publish dozens of such journals, often devoted to a small phase of a discipline. Large societies may have over 50,000 members and thus will have enough support to sustain so many titles; membership in the societies usually includes one or more subscriptions as part of the cost of annual dues. An example of a journal published by a professional society is the highly regarded *Journal of the American Society of Civil Engineers.*

Some professional journals consist exclusively of short articles and letters, perhaps no more than 300 words in length. They are geared to quick publication so that information about important sci-tech discoveries in particular fields can appear in print as soon as possible. Inclusion of contributions of this sort are features of two very prestigious journals, *Science* (published in the United States) and *Nature* (published in England). They are renowned for their letters and short articles, which appear in the same issue as conventional length articles. Many of the most important sci-tech developments are reported in these two journals.

Abstracting and Indexing Services

The most efficient way to locate information contained in journals is to use abstracting and indexing services, many of which have three formats: printed versions, online versions, and CD-ROM versions. There are scores of printed indexes; each has its strong points. For example, some have detailed annotations of each article, as exemplified by *Chemical Abstracts,* which regularly indexes more than 14,000 scientific and engineering journals. Indexes, providing access by authors, keywords, and patent numbers, are issued weekly, cumulating every six months. These cumulations provide other means of access, as do quinquennial cumulations. *Chemical Abstracts* is one of the few services that regularly index chemical patents; its coverage is international in scope. It is available online and in a CD-ROM format.

Other indexing services are devoted to broader fields, such as *Applied Science and Technology Index,* which aims at covering all fields of scientific and technical applications. Unlike *Chemical Abstracts,* it has no annotations of articles. It is restricted to English-language journals that have been selected as being of most interest to libraries of all sizes. It has an online and a CD-ROM version. An indexing service that deals with the health sciences, along with certain aspects of chemistry and biology, is *Index Medicus.* Along with its online counterpart, MEDLINE, *Index Medicus* covers hundreds of periodicals.

In contrast to the above examples, there are abstracting and indexing services devoted to one limited subject field, such as *Mineralogical Abstracts,* which is confined in scope to literature on minerals. It should be noted that many of these services frequently index books and conference papers as well as periodical articles.

A related type of reference tool is the directory of periodical titles, giving names, addresses, and subject coverage of titles, usually on an international basis. Several commercial publications of this type are available. Another useful publication is the union list of serials, which identifies libraries holding any of thousands of titles; this sort of tool can result in saving a great deal of time when seeking the source for a particular periodical. A notable example is *Chemical Abstracts Service Source Index (CASSI),* which provides a listing of the thousands of journals indexed by that service and in which collections of participating sci-tech libraries they may be found.

Costs of Journals

One major problem with sci-tech journals is their high cost relative to serials devoted to other subjects. The largest part of any sci-tech library collection budget is for periodical subscriptions, amounting to at least 70% in some academic sci-tech libraries. A study of the annual survey of subscription costs published in *Library Journal* regularly shows sci-tech journals to be the most expensive titles. For example the article covering 1990 subscription prices showed the following costs for various types of periodicals:[3]

	1988	1989	1990
Soviet Translations	$592.22	$621.70	$678.09
Chemistry & Physics	$329.99	$367.99	$412.66
Engineering	$114.83	$128.37	$138.84
Medicine	$180.67	$199.22	$217.87
Sociology & Anthropology	$64.27	$66.73	$77.61
Law	$43.33	$46.01	$50.32
Education	$47.95	$51.43	$56.33
Literature & Language	$28.04	$29.41	$30.63
General Interest Titles	$28.29	$29.69	$31.24

This table shows clearly how expensive sci-tech journal subscriptions are when compared to those of other disciplines. The 1990 costs represented an increase over 1989 prices of around 12% for chemistry-physics subscriptions; the upward trend has affected prices for several decades. About the only occasion for general decreases in costs, which occur rarely, is lower prices for foreign periodicals when the rate of exchange happens to favor the United States.

TYPICAL EXAMPLES OF JOURNALS

Journal for the Layperson

Scientific American. New York: Scientific American; 1845– .
This periodical has won favor with a wide audience, but it is aimed especially at the lay audience, those who are interested in science and technology but may have limited formal training in those disciplines. Its articles are clearly written and extremely well illustrated, covering a multitude of topics. It also includes book reviews.

Journal for a Specific Discipline

American Society of Civil Engineers. Proceedings. New York: American Society of Civil Engineers; 1873– . Monthly.
One of the older journals for engineers, specifically civil engineers. All of its articles are concerned with some aspect of this branch of engineering, whether road building, design of bridges, or new materials for construction projects. It is typical of hundreds of journals that are published by professional societies for the benefit of their members. Besides regular articles there are also some news items about the profession.

Journal for the Professional in Many Fields

Science. Washington: American Association for the Advancement of Science; 1880– . Weekly.
Combines articles with short capsule summaries of research in many disciplines. The short summaries make this journal very much the favorite medium in the U.S. for announcements of recent developments in science; it is probably the one journal in the U.S. that is most apt to carry the first public disclosure of important new research results. Most of the contents would be meaningless to the layperson; it is aimed at scientists. There are occasional articles on broader topics, such as the relationship of science to government. Includes book reviews.

TYPICAL EXAMPLES OF BIBLIOGRAPHIC CITATION SOURCES FOR JOURNALS AND ARTICLES

Abstracting Service for a Few Major Disciplines

Chemical Abstracts. Easton, PA: American Chemical Society; 1907– . Weekly.
A highly respected indexing and abstracting service known for its detailed, informative abstracts. Chiefly covers chemistry, chemical engineering, and biochemistry. Provides coverage of more than 14,000 periodicals, as well as patents and selected conferences, on an interna-

tional basis. Has indexes issued on several time schedules: weekly, semi-annually, annually, and every five years (formerly every 10 years). Access is by authors, subjects, patent numbers, molecular formulas, organic structure, and chemical substance numbers. Also available online.

Indexing Service for Many Disciplines

Applied Science and Technology Index. New York: H. W. Wilson Company; 1913– . Monthly (except August).
Aims at very selective coverage of all areas of science and technology; restricted to less than 300 English-language periodicals. It has no annotations and allows for retrieval only by subjects. There is an index of reviews of selected new books on sci-tech subjects. Also available in recent years in online and CD-ROM formats.

Indexing Service for a Few Disciplines

Index Medicus. Washington, DC: National Library of Medicine; 1960– . Monthly.
Indexes hundreds of periodicals in the health sciences. Provides access by authors and subjects. Its online counterpart is MEDLINE.

Abstracting Service for One Discipline

Mineralogical Abstracts. London: Mineralogical Society of Great Britain; 1920– . Quarterly.
Restricted to abstracting books and journals on mineralogy, on an international basis. Has an arrangement of entries by subjects, with annual topographical, author, and subject indexes.

Union List of Serials

Chemical abstracts service source index (CASSI). Columbus, OH: Chemical Abstracts Service; 1975– . Quarterly with annual cumulations.
Presents bibliographic information for tens of thousands of publications indexed by *Chemical Abstracts* and related to chemistry, biology, chemical engineering, and other disciplines. Also available online. In 1990 a cumulative printed index covering 1907–1989 was issued, listing data on 68,000 serials, conference proceedings, and other nonserials, along with holdings of 350 participating sci-tech libraries in 28 countries.

REFERENCES

1. Subramanyam, Krishna. *Scientific and technical information resources.* New York: Dekker; 1981: p. 30–31.

Devotes some 30 pages to a review of the history, characteristics, and signficance of journals. The problems and future outlook for journals are also covered. Provides a balanced, thorough discussion of the subject.

2. De Solla Price, Derek J. *Little science, big science.* New York: Columbia University Press; 1963: p. 8–9.

Includes a chapter that provides a fascinating review of the development of sci-tech journals.

3. Young, Peter R.; Carpenter, Kathryn Hammell. Price index for 1990: U.S. periodicals. *Library Journal.* 115(7): 50–56; 1990 April 15.

Gives statistics on journal costs in all disciplines. Generally considered a reliable guide to periodical costs and trends. A companion article covers costs of serial services (abstracting and indexing services).

Chapter 6
Laboratory Notebooks

HIGHLIGHTS OF LABORATORY NOTEBOOKS

Description: Laboratory notebooks consist of bound books in which scientists record events taking place in their laboratories during the course of research. They constitute a record that can be used to prove that certain events or discoveries took place in a given laboratory at a particular time.

Significance: Laboratory notebooks are indispensable documents for use by organizations in establishing that certain scientific discoveries occurred on a particular date in the organization's laboratories. This can be invaluable in pursuing new patents involving the discoveries or in protecting old patents. The techniques described in the notebooks can also be useful in instructing new researchers.

Quantity: There are tens of thousands of laboratory notebooks in existence; practically every research organization will have an array of such records covering its years of research.

Physical Characteristics: Laboratory notebooks are usually small paperback books consisting of numbered, lined pages. Each book bears a distinctive number.

Availability: Laboratory notebooks are for use only within an organization and then only by people authorized to see them.

Retrieval: Each organization devises its own method of retrieval. It could be a very crude method, more or less dependent upon knowing the name(s) of the researcher(s) involved. Alternatively, more sophisticated methods could be used, such as a computerized system allowing for searching by subjects and other means of identification.

Intended Audience: Authorized readers include an organization's scientists, lawyers, and executives.

Scope: Laboratory notebooks are very narrow in scope. A typical notebook consists solely of a detailed record of the experiments of one scientist for a period of years.

NATURE OF LABORATORY NOTEBOOKS

It is very important in the scientific world to have accurate records of what occurred during the course of research in laboratories. This evidence can aid an organization in obtaining new patents or in protecting its right to previously granted patents. Another value of such data is their use in training new researchers, who can study the notebooks as a guide to correct laboratory procedures.

Most research organizations routinely require their scientists to record daily events in the laboratory, using the kind of notebook chosen by the organization and following guidelines established by the organization. A record is kept of the number of the notebook and the person to whom assigned. Most organizations use the company library for recording the notebook assignments and for establishing some sort of retrieval system.

An article by C. Margaret Bell shows that retrieval systems used by various companies range from a simple card file showing the number(s) of the notebook(s) assigned to a particular scientist to an elaborate online system that allows for retrieval by name, subject, project number, ingredients used, and so on.[1] One company has even indexed the notebooks by sections of 10 pages each, thus making it possible to find specific details of laboratory research.

With the advent of microcomputers and many types of software for building an online record, there is no reason why any organization cannot have a carefully indexed, easily retrieved set of notebooks.

REFERENCE

1. Bell, C. Margaret. Laboratory notebook storage and retrieval systems. *Science & Technology Libraries.* 1 (4): 65–71; 1981 Summer.
Describes the findings of a survey of the methods used by 83 food science libraries in providing retrieval of their respective laboratory notebooks. Four different systems in general use are discussed.

Chapter 7
Monographs

HIGHLIGHTS OF MONOGRAPHS

Description: Monographs are traditionally defined as books dealing with a specialized topic, written for people with some degree of knowledge of the subject, and not usually designed for use as textbooks. A large, comprehensive monograph is sometimes called a treatise.

Significance: Monographs serve the purpose of presenting a sizable amount of information on a limited topic to readers with a background in the subject. A monograph does not cover as much ground as a textbook, but it usually treats a particular subject in greater detail than a book intended for classroom use. Monographs frequently include data of a research nature.

Quantity: There are thousands of sci-tech monographs written each year. Of the approximately 10,000 new sci-tech books published annually, a high percentage are monographs.

Physical Characteristics: Physically, monographs have few distinctive features in common. Some are hardbound; some are softbound. Some are attractively printed and bound; some are not. Most monographs have indexes.

Availability: Many monographs are to be found in appropriate library collections, but some are more easily located than others, depending upon their popularity. The more esoteric the subject, the greater the likelihood that only a very specialized library will own a copy.

Retrieval: There are no special techniques for retrieval of monographs, since most catalogs (whether online or in card form) allow for searching by author, title, subject, and series. There are special publications and computerized files that list both monographs and textbooks, but there is rarely a way to search for either monographs or textbooks exclusively.

Intended Audience: As previously noted, monographs are designed for readers with some background in the subject. This allows for a wide range of difficulty of comprehension. Some monographs are understandable only to graduate students or experienced practitioners in the field. Others can be understood by a well-informed layperson.

Scope: A monograph emphasizes a portion of a larger field. For example, a monograph would not attempt to cover all of electrical engineering or all of biology. Rather, one could find a monograph on such specific topics as the design of electric motors or on techniques of genetic experimentation.

NATURE OF MONOGRAPHS

Monographs deal with specialized topics. A book on the design of suspension bridges might be a suitable monograph topic; a book dealing with all aspects of civil engineering would undoubtedly be written as a textbook. (*See* Chapter 31, Textbooks)

Many monographs are written by outstanding scientists and engineers for an audience of their intellectual peers; the level of the text is not diluted or simplified. Sometimes a monograph is the first publication to report on certain new developments, usually of a very specialized nature. Revised editions are sometimes published, particularly in the case of books written by the top people in various disciplines. While books by such authors are authoritative and frequently have a long life, they are not designed for quick use and hence would be unlikely candidates for a reference collection in a library. They require hours of study to master, are not designed for casual browsing, and need to be available for prolonged time periods for the average reader.

There are some exceptions to the general rule that monographs are not used as textbooks; the further one goes up the academic ladder, the greater the possibility that advanced courses or graduate school courses in general might use certain monographs as texts. One reason for this is the ability of advanced students to comprehend the subject matter without the special treatment given difficult material in most textbooks.

Traditionally, the information in monographs has not been as readily available as that in periodical and journal literature. Few indexing or abstracting services analyze the contents of monographs in any depth, and the primary access to the subject matter has been through broad subjects listed in publishers' catalogs or bibliographies. One well-known tool for locating books by titles, authors, and subjects is *Books in Print*, which is now available in printed, online, and CD-ROM versions. The CD-ROM version allows for more complex subject searches, as well as retrieval by means of language of publication, educational level, date, publisher, and a variety of other means of access. For many years there has also been an abridged version of the printed index known as *Scientific and Technical Books and Serials in Print*. As the title indicates, it is limited to books dealing with science and technology, along with a large listing of sci-tech periodicals.

Authors often identify their books as being monographs or textbooks; such information, if available, is an aid to those attempting to decide upon purchase of sci-tech books.

TYPICAL EXAMPLES OF TOOLS FOR LOCATING MONOGRAPHS

BOOKS IN PRINT PLUS. 4th ed. New York: Bowker; 1989. CD-ROM with user's manual.

This CD-ROM product, designed to be used with MS-DOS computers, provides information on over 840,000 titles in the *Books in Print* database. Searchable fields include author, title, key word, publisher, subject, ISBN number, Library of Congress card number, series, audience for whom intended, grade level, illustration, language in which written, price, year of publication, and author/title combinations. A separate product, entitled BOOKS IN PRINT WITH BOOK REVIEWS PLUS, adds reviews from eight journals, including *Sci-Tech Book News*. *Books Out-of-Print Plus* provides publisher survey information about titles declared out-of-print or out-of-stock by many publishers.

Scientific and Technical Books and Serials in Print. New York: Bowker; 1971– . Annual.

Limited to those books listed in *Books in Print* which deal with science or technology, it also has a large listing of sci-tech periodicals and journals, representing a global selection. Several types of indexes are provided, such as titles and authors.

TYPICAL EXAMPLES OF MONOGRAPHS

Carroll, Raymond J. and Ruppert, David. *Transformation and weighting in regression*. New York: Chapman and Hall; 1988. 249 p. (Monographs on Statistics and Applied Probability Series).

States clearly in the preface that the book is a monograph, not a textbook. The book concentrates on the analysis of regression data in certain special cases, summarizing the techniques found useful by the authors for problems typically encountered. Clearly not written for newcomers to the field.

Fites, Philip and others. *The computer virus crisis*. New York: Van Nostrand Reinhold; 1988. 171 p.

Contains technical information on specific viruses as well as general data on the whole phenomenon. Describes what computer engineers recommend for detecting a virus as well as measures for preventing their spread.

Chapter 8
Patents

HIGHLIGHTS OF PATENTS

Description: A patent is a right granted by a government that gives a person or organization exclusive ownership of the design of a particular product or of the method of performing a particular process. In the United States a patent is good for 17 years and cannot be renewed.

Significance: Patents can be very valuable if they cover a product or process that becomes popular in the commercial marketplace. On the other hand, patents can lead to expensive lawsuits if two parties get involved in a dispute over the scope of a patent and its illegal use.

Quantity: There are millions of patents in existence; almost every industrialized nation in the world has its own patent system. In the United States more than 4 million patents have been issued since 1789.

Physical Characteristics: A typical patent is a document with a drawing of the product displayed on the cover, followed by a descriptive text. Figure 8-1 displays one of the pages of drawings from U.S. patent 3,422,476 for a machine used in manufacturing shoes. Figure 8-2 displays the abstract and description of the various parts of the same machine. In recent decades patents have become available on microfiche, microform reels, and, more recently, in computerized versions.

Availability: Patent documents are readily available for sale by the U.S. Patent and Trademark Office, or may be read without cost at any of the dozens of patent depository libraries in this country. A few of the larger depositories may also collect foreign patents. Private corporate libraries, usually not open to the public, often have extensive holdings of foreign patents, generally concentrated on some specific topic. There are organizations that specialize in selling specific patents, often on a rush basis, if desired.

Figure 8-1. Patent Drawing

Jan. 21, 1969   M. M. BECKA   3,422,476

METHOD AND APPARATUS FOR CLAMPING AN END OF A SHOE ASSEMBLY

Filed April 19, 1967   Sheet _2_ of 8

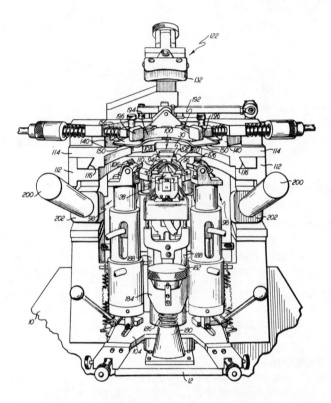

Retrieval: One of the best ways to locate a particular patent is by searching online databases devoted to patents. A manual search of printed abstracting and indexing services not only takes longer but may not retrieve what is wanted due to limitations of the printed services.

Intended Audience: Most patents are written in technical language that cannot be readily understood by the general reader. They are written for those with a strong technical or scientific background, not for the layperson.

Scope: Patents are invariably quite narrow in scope, applying only to one particular product or process, or, more commonly, to an isolated part of a product or process.

Figure 8-2. Patent Abstract and Description

United States Patent Office

3,422,476
Patented Jan. 21, 1969

1

3,422,476
**METHOD AND APPARATUS FOR CLAMPING AN
END OF A SHOE ASSEMBLY**
Michael M. Becka, Cambridge, Mass., assignor to Jacob
S. Kamborian, West Newton, Mass.
Filed Apr. 19, 1967, Ser. No. 632,032
U.S. Cl. 12—145   8 Claims
Int. Cl. A43d 21/00

ABSTRACT OF THE DISCLOSURE

The disclosure is concerned with a machine for stretching a shoe upper about the toe end of the last and then wiping the upper to the toe end of an insole that is located on the last bottom by means of conventional wipers having a flat wiping surface. The machine includes a number of clamping instrumentalities adapted to clamp the shoe assembly in a predetermined configuration with respect to the wipers as well as devices for gripping and stretching the shoe upper about the last.

Summary of the invention

When presenting the toe end of an unlasted shoe assembly to a machine of the instant type having conventional wipers that are adapted to move in a planar wiping stroke the shoe assembly is preferably inclined so that the toe end of the shoe points towards the wiping plane of the wipers. This is done to insure that the extreme toe end of the shoe assembly will be properly wiped to the insole. This is especially true with shoes having relatively pointed toes that are generally more difficult to wipe at the toe. A support is provided in the machine for supporting the shoe assembly in this inclined position. Also included in the machine is a hold-down that is adapted to engage the vamp of the shoe assembly so as to press the forepart of the shoe assembly downwardly against the shoe support and later against the wiping surface of the wipers. A heel clamp is provided that is adapted to preclude heelward movement of the shoe assembly during the wiping operation. When the wiping stroke has progressed to the point where the extreme toe end of the shoe has at least been partially lasted and the wipers have crossed under the insole the support for the shoe assembly is retracted to enable the shoe assembly to be firmly pressed directly against the wiping surface of the wipers by means of the hold down. The inclined orientation of the shoe assembly tends to cause a pitching of the shoe assembly as the support thereof is transferred to the wipers, this pitching tending to cause the heel end of the shoe assembly to move downwardly and become scuffed and marred by reason of engagement with the heel clamp. The instant invention is concerned with a heel clamp, adapted to move heightwise with the heel end of the shoe assembly, so as not to so damage it, yet capable of precluding heelward movement of the shoe assembly during the wiping stroke. It is an improvement over the subject matter disclosed in pending U.S. application Ser. No. 472,525, filed July 16, 1965.

The invention will now be described in greater detail with reference to the accompanying drawings wherein:

FIGURE 1 is a side elevation of the machine, in which the subject of the instant invention is incorporated;

FIGURE 2 is a front elevation of the upper portion of the machine as seen from the line 2—2 of FIGURE 1;

FIGURE 3 is a vertical section of the sleeve and the toe post, movable therein, upon which the shoe assembly is ultimately supported;

FIGURE 4 is a side elevation in section, of the shoe assembly supporting members and the adhesive applicator, that are mounted to the upper end of the toe post and directly support the shoe assembly;

2

FIGURE 5 is a side elevation of the portion of the machine, illustrating the hold down and wiper driving mechanisms and the heel clamp operating mechanism;

FIGURE 6 is a plan view of the wiper driving mechanism taken along the line 6—6 of FIGURE 5;

FIGURE 7 is a plan view of the heel clamp;

FIGURE 8 is an elevation, in section, of the heel clamp as seen along the line 8—8 of FIGURE 7;

FIGURE 9 is a sectional elevation of the mounting for the heel clamp carriage, as seen along the line 9—9 of FIGURE 7;

FIGURE 10 is a front elevation of the toe post extension and the insole rest assembly and adhesive applicator assembly supported thereon;

FIGURE 11 is a plan view of the insole rest assembly and adhesive applicator assembly;

FIGURE 12 is a side elevation, partly in section, and illustrating the shoe assembly just prior to operation of the wiping means;

FIGURE 13 is a plan view of the shoe assembly, illustrating the manner in which the yoke overlaps the peripheral portions of the forepart of the shoe assembly;

FIGURE 14 is a bottom view of the shoe assembly, just prior to actuation of the wiping means taken along the line 14—14 of FIGURE 12;

FIGURE 15 is a side elevation in section of the shoe assembly, as it is initially placed on the adhesive applicator of the machine;

FIGURE 16 is a bottom view of the shoe assembly taken along the line 16—16 of FIGURE 15;

FIGURE 17 is a side elevation, partially in section, of the shoe assembly illustrating the pulling over operation with the bottom of the shoe assembly being brought to bear against the insole rest assembly;

FIGURE 18 is a bottom view of the shoe assembly taken along the line 18—18 of FIGURE 17;

FIGURE 19 is a bottom view of the insole, illustrating the configuration of the ribbon of adhesive applied thereto; and

FIGURE 20 is a side elevation, partly in section, of the shoe assembly and illustrating the position to which the shoe assembly tends to pitch upon retraction of the insole rest assembly.

Referring to FIGURE 1, the machine has a frame 10 and a base plate 12 formed thereon and a sleeve 14 extending downwardly from the base plate 12. For convenience of operation, the base plate 12 is inclined about 30 degrees from the horizontal. For ease of explanation, directions that parallel the plane of the base plate 12 will hereinafter be referred to as extending horizontally and directions paralleling that of the sleeve 14 will hereinafter be referred to as extending vertically. The operator is intended to be located to the left of the machine as seen in FIGURE 1 and a direction extending towards the operator (right to left in FIGURE 1) will be referred to as forward, while a direction extending away from the operator (left to right in FIGURE 1) will be referred to as rearward.

An air operated motor 16 is secured to a cap 18 at the bottom of the sleeve 14, and has a piston rod 20 extending upwardly within the sleeve 14 (see FIGURE 3). A toe post 22 is contained within the sleeve 14 for vertical sliding movement therein and is connected to the piston rod 20 of the motor 16. A roller 24, mounted to the sleeve 14 and extending inwardly thereof, is received in a vertical slot 26 in the post 22 to preclude rotation of the post about the axis of the sleeve 14. The upper end of the toe post 22 extends upwardly beyond the level of the base plate 12. A number of shoe operating instrumentalities, hereinafter described, are supported by the upwardly extending end of the toe post 22 for heightwise movement therewith.

NATURE OF PATENTS

Patents cover a wide range of products and processes. Some are very simple, like a new design for a kitchen can opener, and some are very complex, such as a method for making a delicate microchip containing hundreds of thousands of electronic elements. Most patents are written for specialists, and the expertise required to understand them can be of a very high level. For example, not all chemists would understand all the implications of a chemical patent outside their field of work, nor would all electrical engineers have a firm grasp on the workings of some special device in a different branch of electronics. Patents are practically worthless for people who want to learn about a subject, such as how radar works.

Because patents may bring considerable income to their owners, most organizations require that their employees give up their rights to patents granted on the basis of work done as part of their jobs. In return for the organization's funding for salaries, equipment, advice, and legal counsel, most organizations require that all patent rights be assigned to them. Although the day of the legendary lone inventor, perhaps working in a basement, is not entirely over, more and more patents are granted to employees of organizations, rather than to individuals working on their own.

Patents must represent new developments and must be useful, although patents have been issued for devices that are of questionable usefulness. The application for a patent must include a detailed description of the important features of the design or process; a mere idea is not enough. Patents for devices that attempt to defy the laws of physics, such as perpetual motion devices, are not even considered by government patent examiners.

In recent years questions have arisen as to the legality and wisdom of patenting computer programs and live subjects, such as genetically altered laboratory animals. In 1988 the first biotechnology product, a laboratory mouse, was granted a patent. Computer programs have been patentable for several years now.

Patents have been traced back to the third century B.C. in ancient Greece. But it was not until 1623 that the current patent system with the longest period of continuous operation began, namely that of Great Britain. Actual printing of British patents was not begun until 1852, making them part of the sci-tech literature after that date.[1]

In the American colonies, the colonial governments granted patents until the Constitution gave Congress the power to create a patent system. Thomas Jefferson, as secretary of state, became the first administrator of the patent system. He and George Washington had to sign each patent! Fortunately, this onerous task has long since been assigned to people in less important positions in the government.

The rate of issuance of patents is continually increasing. It took 75 years for the United States to reach the one million mark; then the second and third million came approximately 25 years apart. The four-million mark was reached in 1976, only 15 years later.

Patents provide a unique source of information not always found in conventional literature, such as books or periodical articles. According to several sources cited by Subramanyam, only 5% to 16% of patents are fully described in the so-called open literature.[2] This is contrary to the belief that the technology covered by patents is widely published in books and journals. Therefore, it is vital that there be a clear understanding about ways of retrieving patents.

As mentioned above, there are many printed indexing and abstracting services which include patents appropriate to the subject areas of the respective services. In addition, since 1876 the U.S. Patent and Trademark Office has issued a weekly index, the *Official Gazette*.[3] The *Gazette* may be a good publication to peruse on a regular basis, but it is not a worthwhile source for retrospective searches.

The best sources for such searches are the numerous online databases devoted to patents, examples of which are listed at the end of this chapter. Some are restricted to U.S. patents, while others are international in scope. Like most databases, they can be searched in many ways, such as by patentee, by organizations to whom the patents are transferred, by subject, by date, and by various classification codes used for categorizing the subject matter of patents. Some of the databases are limited in their subject scope, while others encompass all subjects.

TYPICAL EXAMPLES OF DATABASES FOR LOCATING PATENTS

APIPAT. New York: American Petroleum Institute; 1964– .
    Restricted to patents on petroleum products from the United States and eight other countries. Has around 200,000 records.

INPADOC. Vienna: International Patent Documentation Center; 1968– .
    Lists more than 15 million patents from more than 50 countries.

US PATENTS. London: Derwent, Inc.; 1971– .
    Contains coverage of all U.S. patents since 1971. Lists more than 1.3 million records.

REFERENCES

1. Subramanyam, Krishna. *Scientific and technical information resources.* New York: Dekker; 1981; Chapter 6.
Provides a well-written account of the development of patent systems in various areas of the world. Explains many features of patents.

2. Ibid.

3. Ibid.

Chapter 9
Preprints

HIGHLIGHTS OF PREPRINTS

Description: Preprints are copies of manuscripts that have been submitted for formal publication but have not yet been accepted by any publisher. Most often, preprints are unpublished journal articles.

Significance: In some fields of science and technology events move so fast that researchers find the formal process of acceptance and publication of important papers too slow. Distribution of preprints allows for speedier dissemination of information than does formal publication.

Quantity: There are probably several hundred preprints in circulation among closely knit groups of interested scientists at any one time.

Physical Characteristics: Preprints usually consist of photocopies of typewritten manuscript pages. Graphic material prepared to accompany the preprint version of the proposed publication may be rather informal in appearance.

Availability: Most preprints are sent to informal groups of scientists (or to the libraries of the institutions where they work). Individual authors of preprints can always be asked for copies, but it is almost impossible to find out what exists and whom to contact.

Retrieval: There are few satisfactory methods of locating individual preprints because such manuscripts have not yet entered the formal channels of book or journal publication. Little is being done in the way of recording existing preprints; although one serial that attempts this is described later in the chapter.

Intended Audience: Essentially only scientists in certain fields, particularly high-energy physics and particle physics, are apt to be interested in preprints.

Scope: Most preprints have the same scope as journal articles or conference papers, which tend to focus on limited topics.

NATURE OF PREPRINTS

Certain branches of science are developing so rapidly that they produce a constant flow of new theories and new discoveries. It is not surprising, therefore, that researchers in such fields find formal means of publication much too slow to suit their needs. Several months pass from the time the manuscript form of a journal article is submitted until the issue is printed. Papers submitted for acceptance as conference literature can take just as long, or longer, if the conference papers are gathered together and later published as a separate book.

In the past two or three decades preprints have become a common means of disseminating information among researchers in two fields: high-energy physics and particle physics, which use accelerators to investigate the nature of subatomic particles. Researchers in these disciplines need a quick way to be kept informed of new developments.

The informal distribution of preprints leaves something to be desired, since most of them circulate among the friends and colleagues of the authors; at best, copies may be sent to the libraries of universities or institutions where the favored colleagues are employed. Professionals outside the boundaries of what is often called the "invisible college" have little chance of being included in the automatic distribution of new preprints.

Only one source announces the existence and availability of preprints, namely *Preprints in Particles and Fields*, a weekly publication prepared by the particle physics department at Stanford University. It is generally to be found in physics libraries at universities, the same locations in which such preprints are themselves likely to be found. A sample page from this publication is found in Figure 9-1.

EXAMPLE OF A PREPRINT SOURCE

Preprints in Particles and Fields. Stanford, CA: Stanford Linear Accelerator Center Library; 1969– . Weekly.
  Lists new high-energy physics preprints received at the Stanford library. Provides authors, titles, and preprint numbers, with indications as to whether or not the subject matter involves experimentation, theory, instruments, computations, and other categories. A section entitled "Anti-Preprints" consists of a listing of published versions of former preprints.

Figure 9-1. Sample Page from *Preprints in Particles and Fields*

Preprints in Particles and Fields

Report No.	✻	Title	Authors
ILL-TH-89-08	T	DERIVATION OF THE WESS-ZUMINO TERM FOR THE CHIRAL SCHWINGER MODEL USING DIRAC BRACKET FORMALISM. Feb 1989. 6p.	C. Wotzasek (Illinois U., Urbana)
ILL-TH-89-10	T	EQUIVALENCE OF DIRAC QUANTIZATION AND SCHWINGER'S ACTION PRINCIPLE IN THE TWO-DIMENSIONAL MASSLESS ELECTRODYNAMICS.Feb'89.10p.	Clovis Wotzasek (Illinois U., Urbana)
ILL-TH-89-11	T	A STUDY OF THE GENERALIZED SCHWINGER MODEL IN THE HAMILTONIAN FORMULATION. n.d. 22p.	C. Wotzasek (Illinois U., Urbana), C.M. Naon (Stanford U., Phys. Dept.)
ILL-TH-89-12	ET	MULTI - BARYON CONFIGURATIONS IN THE omega STABILIZED SKYRME MODEL. Feb 1989. 16p.	T.S. Walhout (Illinois U., Urbana)
ILL-TH-89-14	T	LATTICE MONOPOLES AND LATTICE FERMIONS. Mar 89 25p.	Simon Hands (Illinois U., Urbana)
ILL-TH-89-17	T	COHERENT STATE PATH INTEGRALS AND THE BOSONIZATION OF CHIRAL FERMIONS. Mar '89.9p.	Michael Stone (Manchester U. & Illinois U., Urbana)
Imperial/TP/88-89/13	T	GEOMETRY OF QUANTUM LOOPS AND KNOT THEORY. Feb 1989. 9p.	M.A. Awada (Imperial Coll., London)
INFN-ISS-88/8	ET	ON SOME LAMBDA AND SIGMA HYPERNUCLEAR STATES. PRODUCTIONS AND THEIR PROPERTIES.Sep'88.20p.	Jan Zofka (Rez, Nucl. Phys. Inst. & Rome, ISS)
INFN-ISS-88/10	ET	CONSEQUENCES OF QUARK STRUCTURE OF NUCLEONS AND FEW NUCLEON SYSTEMS. Dec 1988. 21p.	I.M. Narodetskii (Rome, ISS & Moscow, ITEP)
INFN-ISS-88/14	ET	A y SCALING ANALYSIS OF THE ELECTROMAGNETIC LONGITUDINAL AND TRANSVERSE RESPONSE FUNCTIONS. Dec 1988. 34p.	R. Cenni (INFN, Genoa), C. Ciofi degli Atti (Rome, ISS & Perugia U.), G. Salme (Rome, ISS)
INFN-ISS-88/15	ET	A REALISTIC MANY BODY APPROACH TO QUASIELASTIC (y SCALING) AND DEEP INELASTIC (EMC EFFECT) ELECTRON SCATTERING BY NUCLEI. Dec 1988. 21p.	C. Ciofi degli Atti (Rome, ISS & Perugia U.)
INFN-ISS-89/1	ET	DEUTERON WAVE FUNCTION IN QUARK COMPOUND BAG MODEL. Jan 1989. 18p.	I.L. Grach, Yu. S. Kalashnikova (Moscow,ITEP), I.M. Narodetskii (Moscow, ITEP & Rome, ISS)
KUNS-965	T	NONPOLYNOMIAL CLOSED STRING FIELD THEORY. Mar 1989. 18p.	Taichiro Kugo, Hiroshi Kunimoto, Kazuhiko Suehiro (Kyoto U.)
LA-11491-T	E	MEASUREMENT OF n p ELASTIC SPIN-SPIN CORRELATION PARAMETERS at 484-MEV, 634-MEV AND 788-MEV. Mar 1989. 265p. (Ph.D. Thesis)	Robert William Garnett (New Mexico State U. & Los Alamos)
LBL-26075	I	ELECTROMAGNETIC DETECTORS. Jan 1989. 23p.	Glen R. Lambertson (LBL, Berkeley)
LBL-26415	ET	VECTOR-VECTOR PRODUCTION IN PHOTON-PHOTON INTERACTIONS. Dec 1988. 7p.	Michael T. Ronan (LBL, Berkeley)
LBL-26465	E	RESONANT PRODUCTION IN TWO PHOTON COLLISIONS. Dec 1988. 109p. (Ph.D. Thesis)	Frederick Butler (UC, Berkeley & LBL, Berkeley)
LBL-26561	ET	HIGGS —> FOUR LEPTONS AT THE SSC. Dec'88.10p.	E.M. Wang (LBL, Berkeley), G.G. Hanson (SLAC)
LBL-26593	E	BOSE-EINSTEIN CORRELATIONS OF PIONS IN e+ e- ANNIHILATION AT 29-GeV CENTER-OF-MASS ENERGY. Jan 1989. 98p. (Ph. D. Thesis)	Robert Ellsworth Avery (LBL, Berkeley)
LBL-26802	E	MASS AND TRANSVERSE MOMENTUM DEPENDENCE OF THE DIELECTRON YIELD IN p Be COLLISIONS AT 4.9-GeV. Feb 1989. 12p.	DLS Collaboration (A. Letessier-Selvon, et al.)
LBL-26947	ET	SEARCHING FOR Z-prime AND W-prime BOSONS OF LEFT-RIGHT SYMMETRIC MODELS AT HIGH-ENERGY COLLIDERS. Nov 1988. 11p.	J.A. Grifols, A. Mendez (Barcelona, Autonoma U.), R.Michael Barnett (LBL, Berkeley)
LPTHE-ORSAY-88/34	T	LOGARITHMIC CORRECTION TO THE LOCAL EQUIVALENCE OF PLANAR FIELD THEORIES AND STRING THEORIES. Jun 1988. 19p.	R. Hong Tuan (Orsay, LPTHE)
LPTHE-ORSAY-88/67	ET	NUCLEUS-NUCLEUS COLLISIONS AT CERN: PLASMA OR NO PLASMA? Dec 1988. 27p.	A. Capella (Orsay, LPTHE)
LPTHE-ORSAY-89/01	ET	B MESON DECAY CONSTANT ON THE LATTICE AND RENORMALIZATION. Jan 1989. 38p.	P. Boucaud, Lin Chyi Lung, O. Pene (Orsay, LPTHE)
LPTHE-ORSAY-89/03	T	CHARGE DISTRIBUTION IN SCATTERING OFF NUCLEI: A TEST OF THE PRODUCTION MECHANISM.Jan'89.19p.	A. Capella, C. Merino, J. Tran Thanh Van (Orsay, LPTHE)
LPTHE-ORSAY-89/08	T	BRST APPROACH TO SEMICLASSICAL QUANTIZATION. Feb 1989. 17p.	V. Alessandrini (Orsay, LPTHE)
MAD/PH/481	ET	rho PARAMETER CONSTRAINTS ON FOURTH GENERATION QUARK MASSES. Mar 1989. 10p.	V. Barger, et al.
MIT-CTP-1676	ET	NOVEL CALCULATION OF WEAK MESON NUCLEON COUPLINGS. Jan 1989. 30p.	N. Kaiser (Bohr Inst.), Ulf G. Meissner (MIT, LNS)
MIT-CTP-1678	ET	MICROSCOPICAL STUDY OF THE S WAVE OPTICAL POTENTIAL OF SLOW PIONS SCATTERED BY A NUCLEUS. Dec 1988. 23p.	C. Garcia-Recio (MIT, LNS), E. Oset (Valencia U.)
MIT-CTP-1679	ET	CHIRAL SYMMETRY AND MEDIUM MODIFICATIONS OF NUCLEON PROPERTIES. Dec 1988. 9p.	Ulf G. Meissner (MIT, LNS)
MIT-CTP-1691	ET	QUARK MASS DIFFERENCES AND ISOSPIN VIOLATION IN THE PION - NUCLEON COUPLING CONSTANTS. Jan 89 5p.	Veronique Bernard (Strasbourg, CRN), Ulf G. Meissner (MIT, LNS)
MIT-CTP-1693	ET	HADRONIZATION OF THE FLUX TUBES IN ULTRARELATIVISTIC NUCLEUS-NUCLEUS COLLISIONS.Jan'89.26p.	M. Kataja (MIT, LNS & Jyvaskyla U.), T. Matsui (MIT, LNS)
MPI-PAE/PTh-54/88	ET	FLAVOR CHANGING DECAYS OF THE Z0 BOSON TO MASSIVE QUARKS IN THE TWO HIGGS DOUBLET MODEL. Dec 1988. 16p.	Wei-Shu Hou (Munich, Max Planck Inst.), Robin G. Stuart (Mexico, IPN)
MPI-PAE/PTh-15/89	ET	B-prime MIXING. Mar 1989. 16p.	Wei-Shu Hou (Munich, Max Planck Inst.)
NIKHEF-P-1	ET	THE SPIN STRUCTURE OF THE NUCLEON AND ITS EVOLUTION. Jan 1989. 9p.	J. Kunz, P.J. Mulders, S. Pollock (NIKHEF, Amsterdam)
NORDITA-89/8-P	T	COUPLINGS IN THE HETEROTIC SUPERCONFORMAL THREE GENERATION MODEL. n.d. 34p.	B.R. Greene (Harvard U.), C.A. Lutken (Nordita), G.G. Ross (Oxford U.)
NORDITA-89/11-P	T	NONRENORMALIZATION THEOREM IN BOSONIZED N STRING FORMALISM. n.d. 18p.	E. Nykanen (Nordita)

7 April 1989 ✻ C = COMPUTING E = EXPERIMENTAL I = INSTRUMENTATION

Preprints in Particles and Fields

Report No.	✻	Title	Authors
Print-89-0267 (OREGON STATE)	T	GENERAL SOLUTIONS OF COVARIANT SUPERSTRING EQUATIONS OF MOTION. n.d. 19p.	Corinne A. Manogue (Oregon State U.), Anthony Sudbery (York U., England)
Print-89-0269 (OSAKA CITY)	T	A THEORY OF SPINOR FIELD WITH 'INTRINSIC' ENERGY MOMENTUM P(k) = M (gamma-k) (1-a gamma-5): A PART OF POINCARE GAUGE THEORY OF GRAVITY. n.d. 19p.	Toshiharu Kawai (Osaka City U.)
OUTP-89-12P	T	EFFECTIVE (2+1)-DIMENSIONAL GAUGE THEORY OF SPIN AND CHARGE DEGREES OF FREEDOM WITH CHERN-SIMONS COUPLING. n.d. 12p.	I.J.R. Aitchison, N.E. Mavromatos (Oxford U.)
Print-89-0263	ET	PROTON DECAY AND THE RIP THEORY OF ELEMENTARY PARTICLES. Mar 1989. 19p.	Charles Semiatin (6636 Eberle Dr., Apt. 203, Baltimore, MD 21215)
RAL-89-018	I	DEVELOPMENT OF THE ZEUS CENTRAL TRACKING DETECTOR. Feb 1989. 15p.	ZUES-UK Collaboration (C.B. Brooks, et al.)
RIMS-656	T	REPRESENTATIONS OF U(q)(gl(n,C)) AT q=O AND THE ROBINSON-SHENSTED CORRESPONDENCE. Mar 1989. 27p.	Etsuro Date, Michio Jimbo (Kyoto U.), Tetsuji Miwa (Kyoto U., RIMS)
RUP-89-5	T	COMMENTS ON RENORMALIZABILITY OF OPEN BOSONIC THERMAL STRINGS. Mar 1989. 9p.	Haruo Fujisaki, Koichi Nakagawa (Rikkyo U.)
SISSA-13/88/EP	ET	SUPERSYMMETRY AND RARE B DECAYS. Oct 1988.5p.	F. Borzumati (SISSA, Trieste & INFN, Trieste)
SISSA-151/88/FM	T	MULTILOOP AMPLITUDES FOR THE BOSONIC STRING IN THE OPERATOR FORMALISM. Nov 1988. 14p.	Jorge Russo (SISSA, Trieste)
SLAC-PUB-4558	I	THE MARK-II DETECTOR FOR THE SLC. Mar 1989.91p.	G. Abrams, et al.
SLAC-PUB-4840	ET	STRONG INTERACTION CORRECTIONS TO THE DECAY K —> pi neutrino anti-neutrino FOR LARGE m(t). Mar 1989. 13p.	Claudio O. Dib, Isard Dunietz, Frederick J. Gilman (SLAC)
SLAC-PUB-4852	I	THE MARK-II VERTEX DRIFT CHAMBER. Mar 1989.15p.	J. P. Alexander, et al.
SLAC-PUB-4857	I	STUDY OF STRAW CHAMBER LIFETIME WITH ARGON ETHANE. Mar 1989. 4p.	R. Mir, et al.
SLAC-PUB-4869	ET	PERTURBATIVE QCD ANALYSIS OF THE PION FORM-FACTOR USING A FROZEN COUPLING CONSTANT. 3/89.12p.	Chueng-Ryong Ji (North Carolina State U.), F. Amiri (SLAC)
SLAC-PUB-4884	T	N=2 SUPERCONFORMAL SYMMETRY AND SO(2,1) CURRENT ALGEBRA. Mar 1989. 40p.	Lance J. Dixon, Michael E. Peskin (SLAC), Joseph Lykken (UC, Santa Cruz)
SPI-1490	ET	ON THE EFFICIENCY AND A METHOD OF CHECKING THE RELATIVITY THEORY BASED ON A STUDY OF THE INTERACTION BETWEEN THE HIGH-ENERGY PARTICLES. n.d. 6p.	I. Rozental (Moscow, Space Res. Inst.)
TIFR/TH/88/66	T	STRINGS AND THE RENORMALIZATION GROUP. Jun 1988. 39p.	Sumit R. Das (Tata Inst.)
TIFR/TH/89-8	T	A MEAN FIELD THEORY STUDY OF LATTICE GAUGE THEORY WITH FINITE TEMPERATURE AND WITH FINITE FERMION DENSITY. n.d. 47p.	Satchidananda Naik (Tata Inst.)
TKU-89/01	ET	FINITE W BOSON MASS EFFECT AND BACKGROUND IN NEUTRINO COUNTING REACTION. Mar 1989. 41p.	Masataka Igarashi, et al.
TMUP-HEL-8901	ET	PHOTINO AND SNEUTRINO PRODUCTION e- p —> photino sneutrino q X AT HERA ENERGY. Feb 1989. 10p.	Tadashi Kon, Keiichiro Nakamura, Tetsuro Kobayashi (Tokyo Metropolitan U.)
UICHEP-PUB-89-9	T	BRS COHOMOLOGY IN STRING THEORY: GEOMETRY OF ABELIZATION AND THE QUARTET MECHANISM. Mar 1989. 32p.	Henrik Aratyn (Illinois U., Chicago)
UMDEPP-89-155	T	VIOLATION OF LORENTZ INVARIANCE IN TYPE II GREEN-SCHWARZ SUPERSTRINGS IN CURVED D = 10, N=2 SUPERSPACE. Mar 1989. 11p.	S.James Gates, Jr., Parthasarathi Majumdar, B. Radak, S. Vashakidze (Maryland U.)
UM-P-89/25	T	HORIZONTAL SU(2) SYMMETRY REVISITED. n.d.14p.	R. Foot, et al.
UMTG-148	T	RIEMANN DOUBLE CONSTRUCTION FOR SPINNING STRING PROPAGATORS. Mar 1989. 16p.	Jun Liu, Luca Mezincescu, Rafael I. Nepomechie (Miami U.)
UM-TH-89-05	ET	STRONG CP VIOLATION AND THE NEUTRON ELECTRIC DIPOLE MOMENT. Mar 1989. 12p.	Jiang Liu (Michigan U.), C.Q. Geng, John N. Ng (TRIUMF)
UTHEP-180	T	A PARTICLE PICTURE APPROACH TO INTEGRATED CHIRAL U(1) ANOMALY IN LORENTZIAN SPACE-TIME. n.d. 42p.	Toshiharu Itoh, Kazuhiko Odaka (Tsukuba U.)
UT-KOMABA-89-1	T	EVALUATION OF ONE LOOP MASS SHIFTS IN OPEN SUPERSTRING THEORY. Jan 1989. 35p.	Akihiko Tsuchiya (Tokyo Inst. Tech.), Hisashi Yamamoto, Tsuyoshi Nishioka (Tokyo U., Komaba)
UT-KOMABA-89-2	T	BOSON - FERMION MASS SPLITTINGS IN FOUR-DIMENSIONAL HETEROTIC STRING MODELS WITH ANOMALOUS U(1) GAUGE GROUPS. Feb 1989. 31p.	Masahiro Yamaguchi, Hisashi Yamamoto, Tetsuya Onogi (Tokyo U., Komaba)
UT-KOMABA-89-4	T	ON THE INTERPRETATION OF MINIMAL LENGTH IN STRING THEORIES. Jan 1989. 11p.	Tamiaki Yoneya (Tokyo U., Komaba)
UTTG-06-89	T	DECOUPLING VERSUS EXCLUDED VOLUME OR RETURN OF THE GIANT WORMHOLES. n.d. 24p.	Joseph Polchinski (Texas U.)
UTTG-09-89	T	UNIDEXTEROUS LOCALLY SUPERSYMMETRIC ACTIONS FOR CALABI-YAU COMPACTIFICATIONS.Mar 1989.13p.	S.James Gates, Jr. (Maryland U.), Tristan Hubsch (Texas U.)
UTTG-32-88	T	HETEROTIC (2,2) VACUA: MANIFOLD THEORY AND EXACT RESULTS. n.d. 14p.	Rolf Schimmrigk (Texas U.)
Print-89-0268 (VALENCIA)	ET	NEUTRINOS IN A MEDIUM. n.d. 26p.	Jose F. Nieves (Valencia U.)
Print-89-0265 (VIENNA)	T	CHECKING THE S MATRIX IN QCD IN AXIAL GAUGES WITHIN THE GENERALIZED LM PRESCRIPTION. Mar 1989. 20p.	Gunter Pollak (Vienna, Tech. U.)
WIS-89/11-PH	T	THE RELATION BETWEEN OPERATOR AND PATH INTEGRAL COVARIANT QUANTIZATIONS OF THE GREEN-SCHWARZ SUPERSTRING. Feb 1989. 17p.	E. Nissimov, S. Pacheva, S. Solomon (Weizmann Inst.)

R = REVIEW T = THEORETICAL PPF-89-14

Preprints in Particles and Fields, April 7, 1989. Reprinted with permission of the Stanford Linear Accelerator Center.

Chapter 10
Technical Reports

HIGHLIGHTS OF TECHNICAL REPORTS

Description: Technical reports are documents prepared by organizations to provide information about specific projects or programs of interest to knowledgeable people. They are frequently part of a numbered series.

Significance: Technical reports can be very quickly prepared, should that be necessary. They provide information on detailed subjects without having to include background information in view of the specialized audience for which they are normally intended. Thus they can be concise, providing information that is often difficult to find elsewhere.

Quantity: There are millions of technical reports in existence, counting both government and private sources.

Physical Characteristics: Technical reports tend to be rather short documents, often in paperback format, and may not include much in the way of graphics, such as photographs, drawings, or charts. Some, however, have hard covers, an abundance of artwork, and a very professional appearance. A great deal depends upon the size and type of the intended audience. Since most technical reports are prepared primarily for internal distribution within an organization, they can be informal in appearance, hence less costly to produce and more quickly prepared than literature aimed at a more formal audience. Many technical reports are available on microfiche or microfilm.

Availability: A large percentage of technical reports are not available to the general public because of military or commercial security restrictions; such reports may be obtained only by those qualified to see them, and qualified readers are not free to distribute these documents to nonqualified readers. Unrestricted reports from government agencies are usually sold or distributed without charge to the public. Many can be purchased or obtained at no charge from the issuing agencies.

Retrieval: Adequate access to government technical reports is provided by several indexes; some of these indexes are also available online. Commercial technical reports are usually retrieved within the issuing organization by means of local systems, ranging from simple card indexes to computerized catalogs. Some reports are reviewed in journals, but this is, at best, limited to a small percentage of the total number.

Intended Audience: Most reports are intended for readers who are knowledgeable about the subject matter. These reports are rarely intended to serve as educational tools or as money-making products.

Scope: The typical technical report is confined to rather narrow limits, often describing progress on a particular project; such a progress report could cover a wide range of time, varying from a period of a few months to several years of work. A technical report covering a wide subject scope, such as a discussion on how to build bridges of all sorts, would be most unlikely to exist; a more likely topic for a technical report would be a new way to design the brackets that hold the support cables, or a new way to protect bridge surfaces from effects of corrosion.

NATURE OF TECHNICAL REPORTS

People working on special projects for government agencies or corporations need a medium for keeping interested parties aware of their progress. The medium must allow quick publication of results, incur relatively low production and distribution costs, and be subject to carefully controlled distribution, if so desired. Technical reports are such a medium.

Technical reports are usually concerned with scientific and technical projects, but the term can apply to reports written in other disciplines, such as sociology, finance, and education. A technical report is typically written on a rather narrow topic, is generally aimed at knowledgeable people (rather than the layperson), is often prepared at low cost, and is normally prepared quickly (particularly in comparison with the length of time required to get a book or periodical article published). In this chapter the discussion will be limited to technical reports that deal with science or technology.

The most prolific sources of technical reports are private industry and government agencies. Such reports began to appear in much greater numbers during World War II than during any previous time period. The large number of companies and government agencies engaged in wartime research and development accounted for this rise in the number of technical reports being created. The war required great speed in informing scientists and engineers of recent findings in laboratories or in factories.

Because they are written for specialists who only need to be kept aware of new developments, technical reports are not useful sources for learning basic facts about the topics discussed. Thus such reports have no tutorial aspects, making no attempt to explain points in basic terms. Some reports are issued regularly, such as monthly, semiannual, or annual progress reports, while others may appear as separate documents unrelated to others and issued whenever convenient for the issuing agency.

Because technical reports are not ordinarily given wide distribution by the sources from which they emanate, there is considerable likelihood that a potentially interested reader of a given report might not even know of its existence. This is especially true when compared with the relative ease with which one can learn of the issuance of a new book or the publication of a periodical article. Books and journals are so well covered by announcement tools, advertisements from publishers, and indexing/abstracting services that an alert reader or librarian can rather easily learn what exists on a given topic in book and journal formats. In the case of privately produced technical reports, there is no easy way to become aware of them. For example, a researcher may learn of the existence of the report at a conference, and the library serving the researcher may have no alternative but to apply to the issuing organization for a copy, often using incomplete data in the citation, and hoping for the best.

Restrictions on Technical Report Distribution

In a world in which mutually suspicious nations must co-exist, there is a vital need for protection of information related to the military defense of a nation's people and territories. Perhaps some day weapons and military might will be unnecessary, but until that day comes, military strength will continue to be needed. Maintaining military strength means preventing certain information from becoming known to would-be enemies. Whole departments in the U.S. Department of Defense and related federal agencies are charged with establishing regulations to safeguard certain information and to see that the regulations are followed. Thus a given engineer or scientist may well profit from gaining access to a particular technical report written on a pertinent topic, but unless the individual has what has been called the "need to know" (official authorization for seeing a particular classified document), has received a personal security clearance, also works for an organization authorized to receive such information, and has a secure office or laboratory in which to work with the document, the individual will never gain legal access to the document.

In similar manner the commercial and industrial sectors of most countries have proprietary information to safeguard from competitors. A company that has spent large sums of money and devoted years of research creating a new product or developing a new process is not about to hand that information over to a competitor. Thus commercial security is part of the process of the distribution and the use of many reports; the regulations may differ considerably from those established for military security, but the goals are usually the same. Each company or corporation is responsible for setting up its own means of safeguarding such information. In some companies the information is so valuable that certain departments cannot be entered except by employees of that particular department, and close watch is kept over access to certain documents.

Security regulations, both military and commercial, are part of our society and are generally recognized as impeding the flow of information, but there seems to be no alternative to having them. Efforts to keep military restrictions as lenient as possible without endangering national security have been made, but there is no way that every person's concept of what should and what should not be considered classified information will ever be achieved. A discussion of the types of regulations for military and commercial safeguarding of information may be found in *Top Secret/Trade Secret* by Ellis Mount and Wilda Newman.[1] It also describes how the Freedom of Information Act enables people to gain access to government information that could be safely disclosed.

Government Report Retrieval and Procurement

Fortunately, tens of thousands of documents are not restricted, either as to their announcement or their ready availability. Therefore, useful announcement and access mechanisms exist for such unrestricted reports. One of the most prolific sources of technical reports, the U.S. government, issues a number of excellent indexing/abstracting services, several of which are also available online for quick searches. Some federal agencies, such as the U.S. Government Printing Office, publish documents covering a wide range of topics, many of which are of interest to scientists and engineers. Other agencies, such as the National Technical Information Service (NTIS) in the Department of Commerce, deal almost exclusively with reports about various sci-tech disciplines.

The index produced by NTIS, *Government Reports Announcement and Index*, is an excellent tool for locating reports held by that agency. Searching can be done using authors, agencies, subjects, contract numbers, report numbers, or combinations thereof. Its subject scope is wide, ranging from "automation" to "zirconium"; all aspects of physical sciences, engineering, and management are covered.

Another excellent abstracting and indexing service for technical reports is issued by the National Aeronautics and Space Administration (NASA). Entitled *Scientific and Technical Aerospace Reports*, this carefully produced source provides access not only to NASA reports but also to those from other government agencies, domestic and foreign. Its online version (AEROSPACE DATABASE) includes both report literature and published literature (books and journal articles).

The great quantity of report literature may present formidable problems when selecting and obtaining needed documents, but depository collections in many parts of the United States can save time when a particular document is needed quickly. Also, government and private document delivery agencies generally have some sort of "rush service" for use in emergencies.

In ordering and citing technical reports, it is important to include the various document numbers which are assigned to government-sponsored reports. Including such numbers when ordering documents results in delivery of items much quicker than orders that list only document titles and authors. Two major series are PBs, assigned to documents by the National Technical Information Service, and ADs, assigned by the Department of Defense through its Defense Technical Information Center.

TYPICAL EXAMPLES OF TECHNICAL REPORT SOURCES

Government Reports Announcement and Index. Springfield, VA: National Technical Information Service; 1946– . Semimonthly.
Indexes thousands of technical reports prepared by government agencies or their contractors. It covers a wide range of topics, including engineering, science, and management. It is the best index for coverage of the PB series, one of the main report series in the U.S. Has six indexes for retrieving reports. Also available as an online database.

Scientific and Technical Aerospace Reports. College Park, MD: National Aeronautics and Space Administration; 1963– . Semimonthly.
Primarily indexes NASA reports but includes reports from other government agencies and from other countries. Has indexes by subject, by corporate and personal authors, by contract numbers, and by report numbers. Now available as part of an online database that includes abstracts of published literature. Covers many subject areas besides aviation and astronautics.

U.S. Defense Technical Information Center. *How to get it—a guide to defense-related information resources.* Compiled by Gretchen A. Schlag and Charles E. Reed. Alexandria, VA: The Center; 1989 Jan; AD A201600. 594p.
In one alphabet lists names of organizations, the series codes for their documents, sources for ordering documents, order forms, telephone numbers, etc. Very useful for those searching for or ordering federal technical reports and related documents.

REFERENCE

1. Mount, Ellis; Newman, Wilda B. *Top secret/trade secret: accessing and safeguarding restricted information.* New York: Neal-Schuman Publishers; 1985. 214 p.
Discusses the ways in which restricted documents can be safeguarded by libraries, including both military and commercial restrictions. Information is also given on the workings of the Freedom of Information Act, which aids individuals seeking to obtain federal information that might otherwise be difficult to locate.

PART III

Secondary Sources
of Information
(Textual)

Chapter 11
Almanacs

HIGHLIGHTS OF ALMANACS

Descriptions: Almanacs are compilations of a variety of types of information, often including statistical, astronomical, and meteorological data. This information is commonly arranged according to a calendar. Most almanacs are published annually.

Significance: Almanacs are particularly useful in scientific and technical fields in which seasonal changes are important, such as agriculture, animal husbandry, astronomy, and meteorology.

Quantity: A few hundred almanac titles relating to science or technology are created annually.

Physical Characteristics: Almanacs tend to have paper covers; they are small in size, often printed on inexpensive paper. Since they are published annually, and contain information specifically organized for one particular year, they are generally replaced by a new edition at the end of that calendar year. Some contain color photographs; most have illustrations in black and white.

Availability: Most almanacs can be purchased in bookstores or through subscriptions. General almanacs can be found in most types of libraries. Some specialized almanacs are produced by governmental departments and are available from the U.S. Government Printing Office.

Retrieval: Most catalogs list almanacs as a sub-heading under the name of the topics or as a subject, as well as by title. They are easy to identify.

Intended Audience: Almanacs are intended for any and all individuals interested in the topics covered. Depending upon their contents, they can be equally of interest to the novice and to the expert.

Scope: Some almanacs emphasize such topics as agriculture or weather; others include any information of interest during a calendar year.

NATURE OF ALMANACS

Almanacs are compact reference sources. General almanacs contain information on a variety of topics. Specialized almanacs contain information for a particular field, arranged by day, week, month, and season for one calendar year. Only certain fields in science that are affected by seasonal changes are appropriate subjects for almanacs. For example, almanacs that focus on agriculture provide planting and harvesting information based on a number of factors, especially the weather, that are expected to influence the growth of crops. Meteorological almanacs provide information on weather patterns and trends for anyone interested in the probable weather during the covered year. A typical almanac of this type is *The Weather Almanac*, which both records past weather data and predicts future weather. Astronomical almanacs provide information on the expected movements of stars, planets, and other celestial bodies for the year. *The Nautical Almanac* is a good example of an astronomical almanac.

Data provided in almanacs include statistics, tables, and other miscellaneous information that might be relevant to the year covered. Also, other reference information may be included, such as names and addresses of associations, lists of periodicals or other sources of information, and bibliographies of recommended materials for further reading.

TYPICAL EXAMPLES OF ALMANACS

The Nautical Almanac for the Year. 1989. Washington, DC: Government Printing Office; 1987. Various pagings. Annual.
Identical editions are published in the United States and the United Kingdom. Contents include, by month and day, phases of the moon, calendars, eclipses, planet notes, standard times, star charts, tables for stars and celestial bodies, and other statistical information of importance in navigation.

The Weather Almanac: Edited by James A. Ruffner and Frank E. Bair. Detroit: Gale Research; 1974– . Annual.
Contains all information that might be of interest to someone concerned with the weather, specifically in the United States. Includes predictions, based on past occurrences and records, of storms, severe weather, and geophysical phenomena. Also includes information on air pollution, health-related weather conditions, weather fundamentals, a glossary, and tables of weather in other parts of the world.

Chapter 12
Annuals and Yearbooks

HIGHLIGHTS OF ANNUALS AND YEARBOOKS

Description: Annuals and yearbooks, despite their different names, are almost identical in nature. Both represent summaries of data or events for a particular subject field pertaining to an annual period, typically for the preceding year.

Significance: Annuals and yearbooks enable readers to make comparisons of data among different years and to note significant trends. Publishers of encyclopedias often issue yearbooks to update the original set of volumes.

Quantity: There are scores of sci-tech annuals and yearbooks in existence.

Physical Characteristics: Some annuals and yearbooks chiefly present statistical data in tables and graphs; others resemble monographs or collections of articles.

Availability: Libraries with sci-tech collections usually have some annuals and yearbooks on general topics. Annuals and yearbooks on very specialized topics are likely to be found only in large sci-tech collections.

Retrieval: Most indexes or catalogs have a category for yearbooks and annuals, or at least allow retrieval by titles of such works.

Intended Audience: Yearbooks and annuals are usually prepared for those with some background in a subject, although exceptions exist. A yearbook issued to update an encyclopedia would have a much broader audience than an annual statistical summary for some smaller field of interest, such as statistics for the metal-working industry.

Scope: Annuals and yearbooks prepared for sci-tech applications range from those devoted to a particular field to those that cover all sci-tech disciplines.

NATURE OF ANNUALS AND YEARBOOKS

In every field of science and technology people find it useful to have a summary of the year's developments, whether displayed as statistics or described in text. For example, managers and engineers employed in the manufacturing industry might wish to know the facts about a certain product—the quantity produced, value, sales, exports, and trends in production figures. Multiply this by the large number of separate fields of interest in science and technology and you get some idea of the value of annual summaries of data. An example of this type of publication is *Aerospace Facts and Figures*, which is entirely statistical in nature, providing data about commercial and private aviation. Annuals confined to reporting statistics resemble almanacs, which are covered in Chapter 11.

Annuals prepared for industry—for example, manufacturing, mining, transportation, and construction—are likely to rely on statistical tables to display data. In the sciences, however, annual summaries tend to use more narrative discription. This kind of annual is closely related to another format, the annual review of the literature, which focuses on the year's publications rather than its events. For example, the *Yearbook of Astronomy* presents a summary of current topics of interest without being limited to actual publications describing those findings. It describes the orbits of planets for the year and the phases of the moon, and also outlines upcoming astronomical events. By contrast, an annual review of the literature on astronomy would cite important publications issued during the year and evaluate their significance. Reviews of the literature are covered in Chapter 27.

Yearbooks updating encyclopedias serve a useful purpose—they allow users of the encyclopedia to trace developments for a given year without having to study and evaluate countless journal articles and reports. Naturally, there are practical limits to the number of yearbooks that would be useful. Poring over yearbooks for the past twenty years would not be feasible. Most encyclopedias publish new editions every five years or so, thus putting a limit on the number of yearbooks that would have to be consulted by readers. Probably the best known example of an encyclopedia yearbook is the *McGraw-Hill Yearbook of Science and Technology*. It presents review articles covering major developments, along with brief summaries of key events.

Guides to the literature generally list the most important annuals and yearbooks, but there are undoubtedly many that are little known outside a particular field. One series that is apt to confuse a person investigating the subject is the well-known *Yearbook of Agriculture* series issued by the U.S. Department of Agriculture. It is not made up of yearbooks in the sense described here, since the volumes bear no relationship to developments in a given time period; rather it is a series of monographs, with a different topic chosen each year, such as "Corn" or "Trees."

TYPICAL EXAMPLES OF ANNUALS AND YEARBOOKS

Aerospace Facts and Figures. Washington, DC: Aerospace Industries Association of America; 1945– . Annual.
Reviews important commercial and technical statistics related to commercial and private aviation as well as astronautical events. Contains many tables and charts.

McGraw-Hill Yearbook of Science and Technology. New York: McGraw-Hill; 1962– . Annual.
Serves to update the *McGraw-Hill Encyclopedia of Science and Technology.* Presents articles, arranged alphabetically, which describe the major events of the previous year. May include photographs along with its text.

Yearbook of Astronomy. New York: Norton; 1962– . Annual.
Provides data on astronomical events of the year along with a directory of astronomical societies.

Chapter 13
Bibliographies

HIGHLIGHTS OF BIBLIOGRAPHIES

Description: A bibliography is a collection of citations to sources of information, arranged and selected according to some prescribed rules or standards. The sources may be print or nonprint materials.

Significance: Bibliographies are important time savers for readers seeking information. They are compiled with a particular user or class of users in mind. They can have a wide scope or a narrow one.

Quantity: There are thousands of bibliographies in existence, but they vary tremendously as to scope, date of compilation, types of materials covered, and audience for whom prepared.

Physical Characteristics: Short bibliographies are often incorporated into books, journal articles, and even technical reports; long ones often appear as separate books, sometimes with indexes.

Availability: Access to bibliographies depends on their formats—they are available in the same manner that similar formats of information are obtained. For example, if a bibliography appeared in a periodical, it would be available from the same sources that provide access to periodicals. The same holds for bibliographies appearing as books—use the sources suitable for books.

Retrieval: Most catalogs and indexes are apt to list these publications under the term "bibliographies," used either as a separate heading or as a sub-heading of major terms. Whether one uses a periodical indexing service or a library catalog would depend upon the format in which the bibliography was published (as a periodical article or as a book).

Intended Audience: Bibliographies can appeal to experts in a given field; they are often useful to neophytes or students.

Scope: Bibliographies may be brief or comprehensive, depending upon their intended readership. One bibliography might list the 100 best articles on metallurgy in 1985; another might consist of as few as three citations dealing with electric generators for use in the Arctic.

NATURE OF BIBLIOGRAPHIES

A bibliography can be defined as a collection of information sources centered on a particular topic, arranged in some organized fashion. Bibliographies are very basic reference tools and can appear in many variations of style, scope, and purpose. Some might be aimed at the layperson, while others are useful only to those individuals with some experience in a field. The time span covered could vary from recent months to many centuries. Some are international in scope, while others might be restricted to one language. It is not unheard of for some to include all types of formats (such as journal articles, patents, and technical reports), or, on the other hand, be restricted to one format (such as books or conference papers).

One of the main values of a bibliography is that it saves readers the time and trouble of locating, selecting, and recording the citations themselves. While it would be unusual to find a published bibliography that contains exactly what the reader would have selected personally, it is nevertheless wise when beginning a search to examine existing bibliographies to avoid "reinventing the wheel." Perhaps a suitable bibliography already exists on the subject, thus saving the preparation of a new one.

For many years bibliographies existed only in printed form, as brief listings incorporated into periodical articles, or books, encyclopedia articles, and sometimes as separate volumes. If no suitable bibliography could be found, it could be constructed by examining indexing and abstracting services to compile a list of appropriate entries. Since the 1970s, it has been possible to compile a bibliography by searching online databases. In some cases the databases contain millions of entries, all retrievable in a variety of ways, such as by subject, author, source, or date of publication. They can be searched quickly, and the printing of citations by computer removes much of the tediousness of copying citations by hand that is required in manual searches.

There are several ways in which bibliographies can be arranged. Common methods include arrangement by author, by title, by subject, by date of publication, by format, and by language. Indexes may be created for long bibliographies. For example, a bibliography arranged by authors might have an index to the subject matter of the entries, while a bibliography arranged by subjects might have an index of authors. Sometimes there might be a need for several kinds of indexes. Much depends upon how thorough the bibliography is intended to be and the compiler's estimate of the needs of future readers.

In some bibliographies each entry is accompanied by an annotation or abstract that briefly describes the contents. These summaries are very useful because they give the reader some indication of the probable worth of the cited item. Titles of citations and other bibliographic data, such as authors or dates of publication, aid in identifying a work, but do not give much information as to whether or not the citation would be useful to the reader.

The length and scope of annotations vary widely. A broad summary of the contents of an entry is called an "indicative abstract." A detailed annotation, giving exact information about an entry is called an "informative abstract." Prime examples of informative abstracts are found in *Chemical Abstracts*, the annotations of which are said to be so complete a reader could almost repeat an experiment by reading them.

There are several reasons why some bibliographies do not include annotations. One reason is the extra cost of writing, editing, and printing them. Another is the amount of time required to create them. Informative abstracts, of course, cost the most. Indicative abstracts can be prepared with less expert help, can be written more quickly, and require less editing before publication. In general, the most useful bibliographies have annotations of some sort.

To be truly useful to readers, a bibliography must be kept up-to-date. Many compilers make no effort at updating their bibliographies; others provide supplements for this purpose. Updating can be as frequent as monthly or as infrequent as every five years.

One type of bibliography, exemplified by the well-known *Books in Print* series and its variants, serves the needs of collection developers. One variant, entitled *Scientific and Technical Books and Serials in Print*, is devoted to all types of sci-tech books, while another is concerned only with health sciences books. Some bibliographies of this sort are issued annually, while others are published monthly. Each has its own scope, type of material cited, and intended audience. Most are restricted to books rather than other formats, although a few deal with periodical articles or technical reports.

An example of a bibliography in book form, one not apt to be updated, is the *Core List of Books and Journals in Science and Technology*. It contains approximately 1,500 entries; each item is annotated. By contrast, a bibliography in book form that is designed to be issued annually is *Science and Technology Annual Reference Review*. It is arranged by broad disciplines, then subdivided by formats, such as handbooks or dictionaries.

TYPICAL EXAMPLES OF BIBLIOGRAPHIES

Powell, Russell H.; Powell, James R., Jr. *Core list of books and journals in science and technology.* Phoenix, AZ; Oryx Press; 1987. 144 p.
Describes more than 700 books and approximately 800 periodicals, serving as a selection guide for sci-tech collections. Each item is annotated; descriptions include major topics covered plus possible uses of the book. Covers all areas of science and technology.

Science and Technology Annual Reference Review. Edited by H. Robert Malinowsky. Phoenix, AZ: Oryx Press; 1989– . Annual.
This annual series aims at reviewing selected sci-tech reference publications. The second volume, published in 1990, contains reviews of almost 800 reference sources, such as encyclopedias, dictionaries, and handbooks. There is an evaluative review for each entry, along with complete bibliographic data. Each volume includes a subject index, name index, and title index.

Scientific and Technical Books and Serials in Print. New York: Bowker; 1971– . Annual.
Provides bibliographic and order information for over 120,000 in-print books published or distributed in the U.S., covering all disciplines in the physical and biological sciences as well as technology. Books are arranged by authors in one volume and by subjects and titles in two other volumes. Over 13,000 subject headings are used in the subject index. There is also a title and subject index for some 18,000 serials representing an international listing of journals. A companion set, entitled *Medical and Health Care Books and Serials in Print,* provides similar coverage for books devoted to medicine and health care.

Chapter 14
Biographical Information

HIGHLIGHTS OF BIOGRAPHICAL INFORMATION

Description: Biographical information can be found in a variety of forms, ranging from a brief set of basic facts to long, scholarly accounts. The information can be routine, uncritical, and bland, or highly opinionated and newsworthy in style.

Significance: There are times when obtaining biographical data about scientists and engineers can be of great importance to the inquirer. Business decisions, employment decisions, and even political decisions may depend upon retrieval of accurate, unbiased information about individuals.

Quantity: Hundreds of sources of biographical data exist, varying greatly in scope, number of entries, amount of detail given, and frequency of publication.

Physical Characteristics: By far the largest amount of biographical information is found in printed sources, whether pages in a collected work, newspaper or periodical articles, or complete books. In recent years, however, computerized databases have become important sources of large quantities of data about scientists and engineers.

Availability: Most libraries have some biographical sources. Biographies in book form and the "who's who" type of serials are probably the most common formats. Many popular periodicals also offer such material. The more serious biographical sources are not commonly found in small libraries.

Retrieval: The retrieval techniques used for locating biographical information generally depend upon the format of literature involved; methods vary according to whether the information source used is a journal, book, encyclopedia, or some other format. Searching may involve looking for a particular person's name or for people in certain occupations or fields of research.

Intended Audience: Just as there are many types of sources of bio-graphical data about scientists and engineers, there are many varied classes of users for this sort of information. In addition to employers and colleagues, schoolchildren, college students, business executives, government agencies, and associations commonly require knowledge of the personal and professional backgrounds of those who are or who were involved in technology or the sciences.

Scope: Some sources provide only a few lines of data: name, home address, employment, and perhaps educational background. Other sources devote full pages to a detailed account of a person's career, including professional accomplishments and a bibliography of major publications. Very prominent people may be the subject of in-depth book-length biographies.

NATURE OF BIOGRAPHICAL INFORMATION

Reference librarians are apt to receive inquiries about the lives and achievements of scientists or engineers. The greatest amount of biographical information available concerns people who are leaders in their fields but have not reached the level of such geniuses as New-ton, Edison, or Einstein. Most people realize that internationally known individuals such as these three have been the subjects of a multitude of books and periodical articles. On the other hand, a librarian might be asked to find information on obscure subjects, such as a relatively unknown Belgian chemist who lived in the 1850s.

Fortunately, there are all sorts of biographical sources, some of which are best suited for locating data about well-known individuals and others that are more applicable to searches for people who are not at all prominent. Descriptions of these different types of bio-graphical information sources follow:

Book-Length Biographies. These are usually restricted to nation-ally or internationally known subjects, although occasionally some publisher is willing to devote a book to a less prominent person.

Book-Length Autobiographies. Scientists and engineers rarely take on the task of writing their own biographies, and autobiographies that do get written are rarely the ones that would be of the greatest interest to the average reader. Einstein's own account of his life would certainly have been a gold mine of data; writers attempting to depict his life could never hope to obtain the same information through other sources.

Biographies in Periodicals. These accounts are often concerned with people of less renown than those who are the subjects of books. Depend-ing upon the nature of the periodical, the people described may have been quite obscure until catapulted into temporary prominence. For example, an astronomer who happens to spot some interesting phenomenon during a routine tour of duty at an observatory may be written about in a popular magazine, complete with photographs and interviews, later to fade from public notice, perhaps for the rest of his or her career. Famous people are likely to be written about at almost any time, although anniversaries of births or accomplishments are often the occasion for a writer to choose a particular person. Unlike individuals who are popular for a brief moment, people who win enduring fame tend to be written about during long periods of time, both before and after their demise. Periodicals that deal with current events may include obituaries of well-known scientists and engineers.

Biographies in Newspapers. Newspaper articles generally follow the pattern described for periodicals, although newspapers are even more likely to use timely events as reasons to publish biographical accounts. Sunday supplements are often good sources for commemoration of the anniversaries of important events. Newspaper obituaries are also valuable sources for summaries of the highlights of the careers of famous and near-famous sci-tech personages.

Biographies in Encyclopedias. Many sci-tech and general encyclopedias include biographical sketches, usually restricted to nationally and internationally known figures.

Biographies in Collected Works. There are many directories and biographical dictionaries that contain thousands of biographical sketches, often limited to a few lines for each person. A few collections, such as those issued periodically by certain professional organizations, have a more scholarly perspective. A noteworthy example is the *Biographical Memoirs* published by the National Academy of Sciences. The accounts found in this publication tend to be very complete, with copious citations of the subject's writings.

A number of collected works may deal with all types of scientists and engineers, while others are confined to one narrow discipline, such as electronic engineers. Some collections include sci-tech personages but are not restricted to them, as exemplified by the various "Who's Who" series. A number of these collected works are issued as serials, either on an irregular schedule or perhaps on an annual or biennial basis. The best known sci-tech biographical serial is entitled *American Men and Women of Science*. This prominent and important source is issued biennially and lists over 100,000 subjects. It is also available online. A similar source, equally valuable, is *Who's Who in Technology*, which is updated on an irregular basis; it is restricted to engineers and lists over 35,000 people. By contrast, other publications of collected biographies are issued as monographs that are never intended to be updated.

Festschriften are collections of articles and essays written in honor of a colleague upon some special occasion, such as his or her retirement. These volumes very frequently include a biographical essay.

Membership Directories. The only source of information about thousands of scientists and engineers is the membership directory issued by professional societies, usually on an annual basis. Entries may consist of little more than name, title, and place of employment.

Author Notes in Periodical Articles. Most periodical articles about topics in science and engineering include brief biographical sketches of the authors. The information usually is restricted to name, title, and place of employment. Academic degrees earned and outstanding awards may also be included.

Bibliographies of Biographical Literature. There are a few monographs that attempt to provide indexes of printed biographies, but they are rarely up to date. The *Index to Scientists of the World from Ancient to Modern Times* by Norma Ireland is an example of an older work of this sort; it is invalid for current subjects but would serve well for the period from ancient eras up to the 1960s.

Online Sources. There are at least half a dozen online databases that provide quick access to biographical information about professional level people, some being more oriented to science and technology than others. One of the advantages offered by online sources is the fact that a searcher can combine concepts or terms, such as locating someone in a particular age range who graduated with a certain degree from a specified type of university. Such specifications would be difficult if not impossible to search for using printed sources. The list of examples for this chapter includes one such database, AMERICAN MEN AND WOMEN OF SCIENCE.

Diaries. A rather neglected source of information is the personal diary in which some people record the events of their lives, events that may never be found in any published source. Not many scientists or engineers, particularly in modern times, bother to keep a diary, which is a loss to readers seeking information that lies outside the usual sources. Scholars are the main beneficiaries of diaries.

Personal Interviews. In many instances it is possible to obtain information about a person by interviewing others who knew the subject in some capacity, perhaps as a neighbor, a fellow worker, or an employee or employer. The longer the time span between the period under investigation and the interview, the more likely it is that errors will creep into the accounts, the human memory being what it is. Yet hitherto unknown facts or incidents are frequently brought to light through interviews.

Most discussions of biographical sources include a note of caution regarding accuracy of the data. A biographical sketch written by the subject might not be as accurate as one written by an independent third party. It is only human for people to gloss over their past misfortunes or mistakes in preparing sketches. The more scholarly the publication in which the sketch is published, the more likely it is to be unbiased and thorough in its treatment.

Many of the "Who's Who" collections as well as most membership lists depend upon the subjects to supply the original entry and to update it. The expense of preparing third-party biographical accounts prohibits most such publications from using any other means of collecting data. Thus if extreme accuracy were required, a meticulous searcher would undoubtedly need to corroborate the data found in self-written biographical accounts by means of other sources.

Not every source of biographical information is reliable or worthwhile. A number of publishers solicit for prospects to include in a proposed biographical directory. Such directories are essentially just money-making schemes, with no real purpose other than to enrich the publishers.

A related topic is that of obtaining photographs of scientists and engineers. A few periodicals routinely include a small photograph of each contributor, and the more prestigious collected works may sometimes include photographs. Book-length biographies invariably use photographs to add interest to a text. Families and friends of the subject are often excellent sources of photographs, as are schools and places of employment.

TYPICAL EXAMPLES OF SOURCES OF BIOGRAPHICAL INFORMATION

American Men and Women of Science 1989-1990. 17th ed. New York: Bowker; 1989. 8 vols. Biennial.
This prestigious source of data about more than 125,000 nationally known U.S. and Canadian scientists includes data about location, accomplishments, education, and major writings. One volume is solely devoted to an index of the persons according to their specialties or disciplines. An online version of this index covers data from 1979 to date.

Ireland, Norma Olin. *Index to scientists of the world from ancient to modern times: biographies and portraits.* Boston: Faxon; 1962. 662 p. (Useful reference series no. 90)
Provides an index to books that describe works containing biographical sketches of scientists.

National Academy of Sciences. *Biographical Memoirs.* Washington, DC: The Academy; 1877/79– . Irregular.
Provides detailed biographies of noted scientists, including in-depth citations of the writings by those surveyed. Highly regarded for its scholarly style.

Who's who in technology. 6th ed. Woodbridge, CT: Research Publications; 1989. 2 vols. Updated irregularly.
This relatively new publication lists over 35,000 engineers, including a special listing of those who might qualify as expert witnesses in court cases. Provides such information as education, employment record, and special achievements. Revised on an irregular basis.

Chapter 15
Computerized Information
Sources

HIGHLIGHTS OF COMPUTERIZED INFORMATION SOURCES

Description: Computerized information as a class includes a number of formats, ranging from those which can be easily searched on personal computers in a library or laboratory to those forms which must be processed on professional equipment in a special processing plant. Whatever the physical format involved, computerized information has revolutionized the storage and retrieval of data.

Significance: The sciences and technology have been the leading disciplines to support the development of computerized information. The speed and efficiency with which information can be retrieved using computerized sources is especially welcome in the sci-tech world because of the huge quantity of data it generates.

Quantity: There are thousands of computerized database files, hundreds of different products designed to be used on laser disks, and thousands of engineers creating new files and new products for use on computers. No other disciplines have as many computerized sources of information as do the sci-tech areas.

Physical Characteristics: Computerized information is usually stored in magnetic or optical formats. In the case of magnetic formats, the sources range from the familiar diskettes used in personal computers to large disks designed for main frame computers. Optical storage is commonly found in the form of CD-ROM disks, although other variations may be used for some applications.

Availability: Computerized searching can now be carried on in practically every corner of the globe, thanks to the extension of communication networks by satellites beyond locations served by telephone landlines. The necessary equipment for searching online databases includes a terminal, a modem, and a printer, all available at a large range of prices. CD-ROMs require a computer and a special disk player.

Retrieval: Searching online computerized information sources requires the proper equipment supported by sound training in its operation. Some systems are more difficult to learn than others. Whatever the requirements, computerized information sources generally provide quicker results than manual searching. They have the further advantage of allowing for searching of complicated relationships among the data sought that would not be possible with manual searching.

Intended Audience: Although the primary users of computerized databases are adults, particularly those with some training, younger people are also being introduced to computerized searching in certain areas. Some databases require a strong background in sci-tech disciplines to be properly used, while others make less stringent demands on searchers. CD-ROM devices are particularly well suited for use by those with a minimum of sci-tech training and instruction in retrieval techniques.

Scope: There are scores of databases devoted to sci-tech disciplines; some are limited to one subject field, such as the earth sciences, while others cover all areas of science and technology. A growing number of these databases are now available as CD-ROM products, thus broadening the number of potential searchers. Some sources are general enough to be used by lay people, while others are so sophisticated that a strong background in the discipline involved is required.

NATURE OF COMPUTERIZED INFORMATION SOURCES

During the past 20 years a major change has taken place in the ways in which information can be stored and retrieved. The creation of stores of data in computerized format has in many ways revolutionized the traditional ways of retaining, editing, sorting, and locating desired information. While this revolution has occurred in all sorts of disciplines—the humanities, social sciences, and other fields—it is particularly well advanced in the sciences and engineering.

The first computerized products were developed in the 1970s and tended to be devoted to science and engineering. While other disciplines now share this type of format, a high percentage of computerized information is still of a sci-tech nature. The first type of computerized source of any real consequence was the online database, available to all searchers who had access to the computer containing the file. This type of service could not have been economically feasible until advances in communication and in computers reached a level of sophistication to allow it to be offered. For example, only when time sharing and inexpensive telecommunication links were fully developed did it become feasible for searchers in many locations to search a given database at the same time, thus lowering costs to affordable levels. The next requirement was for the appearance of database vendors, who lease rights from the various creators of different databases to mount their computerized files in the vendors' systems. The vendors then widely advertized the availability of the service, thus giving the databases a wider audience than the creators could have expected to provide by themselves. The emergence of competing vendors gave database owners an opportunity for either wider exposure in more than one vendor's system or else a more profitable contract with one vendor. Today there are dozens of vendors, some offering hundreds of different databases to potential customers. Access requires only a password, a personal computer, and the communication software and hardware necessary for a link to the vendor.

The contents and purposes of databases are quite varied; some are merely online versions matching the printed products. For example, COMPENDEX PLUS is the name for the database that matches *Engineering Index.* Others contain data not found in the printed indexes. An example of this is *Chemical Abstracts.* By using the online version of the renowned printed index, it is possible to construct online graphic reproductions of the structures of molecules, something not possible from use of the printed version alone. Some databases cover a wide range of topics, such as all of science and engineering; SCISEARCH is an example of such a wide ranging database. Others are restricted to one field, such as POLLUTION ABSTRACTS, or CLAIMS/U.S. PATENT ABSTRACTS, which consists of the claims found in U.S. patents for recent decades. Some databases provide access to all types of data sources, such as books, periodical articles, patents, technical reports, and conference papers, while others may be restricted to only some of these types. For example, the NTIS database consists almost entirely of technical reports handled by the National Technical Information Service, a government agency responsible for the indexing and sale of government-sponsored research reports.

Some databases relate to a single aspect of a given field, such as environmental engineering problems, while other databases might be concerned with engineering as a whole. Some are restricted to English-language materials, while others may be multilingual in scope. Some contain annotations or abstracts of the items cited, while others do not, and still others give the full text of the material that is indexed. Obviously, selection of the proper database to search is critical. Knowing how to do a search correctly and efficiently is itself a skill that requires training and practice.

Online databases have become enormously popular, one reason being the speed with which databases, even those containing millions of records, can be searched. Furthermore searches can be quite complicated as to the number and the relationship of the kinds of terms being sought, such as combinations of subjects, names of authors, and dates of publication. The amount of information that can be searched is usually far greater than that contained within the walls of the library or organization doing the search. The resulting search can be printed by the attached printer in various optional degrees of completeness, thus eliminating the chore of manually copying pertinent data.

An example of a complete full-text database is DRUG INFORMATION FULLTEXT. It provides full-text evaluative information from two publications of the American Society of Hospital Pharmacists, and covers approximately 60,000 marketed U.S. drugs as of 1987.

Some databases consist essentially of numerical data, rather than citations in bibliographic format. An example of a scientific non-bibliographic database is CHEMICAL EXPOSURE produced by the Chemical Effects Information Center at the Oak Ridge National Laboratory. It contains data on chemicals in tissues and body fluids of humans and animals; it also identifies "body burdens which reflect exposure to contaminants in air, food and water as well as to the administration of pharmaceuticals." From data collected since 1974, subject coverage includes chemical identification and chemical properties, analysis methods, toxicology, health effects, pathology, demography, and notes on experimental studies.

The advent of online databases has not eliminated the printed abstracting and indexing service for several reasons, one being that not every search is complicated enough to require the initiation of a computer search. Not all clients like to turn their searches over to surrogate searchers nor do they want to do the searching themselves. In some cases the cost of online searches may not warrant their use, making a manual search necessary. Finally, not all the useful data in the sci-tech world is ever going to be in machine-readable form, such as certain old and little-used information, making manual searches necessary in order to retrieve it. COMPENDEX PLUS, for example, begins with records first produced for *Engineering Index* in 1970. Because of the low use of old engineering data, it is unlikely that material predating 1970 will ever be added to this database.

In the past two or three years a new format has become popular for retrieving information. It is called the CD-ROM, which stands for Compact Disk, Read Only Memory. Essentially it is a computer diskette on which data is permanently recorded. Unlike most computer diskettes, which can be added to, edited, or erased, the CD-ROM remains as prepared, just like the traditional phonograph record. Equipment to make it possible to write on or erase such disks is available, but the majority of products remain as manufactured. CD-ROMs require a special player, which uses a laser type reading device to search data and display it on a computer monitor. Like databases, CD-ROMs can be searched in random fashion, not in sequential fashion, which differentiates the CD-ROM from a conventional phonograph record.

One of the main advantages of a CD-ROM disk is that once it is purchased or leased, the user can search it as much as desired with no charge. With online databases, the user is charged according to such factors as length of time online, cost structures of the databases used, and type of records retrieved (brief citations being cheaper to retrieve than complete records). However, unless a library has obtained a special device that allows for several disks to be operating simultaneously, only one user can search a CD-ROM at a time, whereas a busy library could have more than one password, if desired, so that more than one online search could be carried on in a given library at the same time.

Another disadvantage of CD-ROMs is that they cannot contain information as recent as an online database could provide; most CD-ROM titles offer updated supplementary disks no more frequently than every quarter. Many users feel the CD-ROM is better for inexpensive searching of older material, with online databases reserved for searches requiring the most recent data. Some of the CD-ROMs present combinations of files and records otherwise issued separately as online and printed files. A notable example is a CD-ROM entitled SCITECH REFERENCE PLUS. It consists of five databases on one disk, ranging from sci-tech portions of *Books in Print* to *American Men and Women of Science*.

Some of the most specific information about what is available online is found in the catalogs prepared by the various vendors of databases. However, there are printed directories offering this information; *Index and Abstract Directory* indicates all the databases that regularly index a given periodical and also provides many details about the periodicals themselves. Printed directories of what CD-ROM products exist have also appeared.

COMPUTER SOFTWARE

The computer software which is required to govern the operation of the hardware is, of course, very important. Its quality determines the speed, convenience, and effectiveness of operations. Creators of software for sci-tech purposes must possess considerable training in the disciplines involved in order to understand what the software must accomplish.

Directories of such software exist in various formats, usually confined to products that are available from the originating organization or are for sale from commercial sources. One example is the *Directory of Computer Software, 1988*, published by the National Technical Information Service. It describes over 1,700 software programs for use in science and technology projects. Still another index is issued by Engineering Information, Inc., entitled *Engineering & Industrial Software Directory*. It indexes hundreds of programs; a sample citation is shown in Figure 15-1. Full descriptions are given of uses and operating requirements. The creation of computer software is a fast-moving field, and online databases exist which serve to update existing printed indexes, such as Bowker's MICROCOMPUTER SOFTWARE AND HARDWARE GUIDE. A recent paper by Beth Paskoff summarizes the selection and applications of software in libraries.[1]

TYPICAL EXAMPLES OF COMPUTERIZED INFORMATION SOURCES

Databases

COMPENDEX PLUS. New York: Engineering Information, Inc.; 1970– . Updated monthly.
  A database consisting of the online version of *Engineering Index* that indexes approximately 4,500 journals on an international basis. Subjects include all types of engineering, geology, management, and applied sciences. Also indexes many conference proceedings.

NTIS. Springfield, VA: National Technical Information Service; 1964– . Updated biweekly. Various paging.
  Containing nearly 1.5 million records, mostly unclassified technical reports prepared by the federal government and its contractors, this database is closely related to the contents of *Government Reports Announcement and Index*, a printed abstracting and indexing service.

SCISEARCH. Philadelphia: Institute for Scientific Information; 1974– . Updated biweekly.
  This database provides a multidisciplinary approach to all areas of science and technology. Contains all the records in *Science Citation Index*, amounting to more than 10 million items. Also allows for citation indexing (searching of references at the end of articles).

SCITECH REFERENCE PLUS. New York: Bowker; 1990– . Updated annually.
  Offers a package of five files on one CD-ROM disk, consisting of sci-tech citations from *Books in Print, Ulrich's International Periodicals Directory, American Men and Women of Science, Directory of American Research & Technology*, and *Corporate Technology Directory*. Provides access to over 120,000 books, 17,000 periodicals, 140,000 names of scientists, 11,000 research facilities, and 17,000 corporate research centers.

Figure 15-1. *Engineering & Industrial Software Directory* **Sample Citations**

PROGRAM NAME	**TRI*THERM™ (Heat Exchanger Thermal Design)**
DESCRIPTION OF PROGRAM	TRI*THERM™ is a user-oriented program for the thermal design and rating of heat exchangers. TRI*THERM™ has its own properties databank and VLE routines. It is used by companies in the petrochemical, refining, fabricating, power and gas processing industries. The program has a built in databank that contains...
OPERATING ENVIRONMENT	**Computer Hardware:** CDC Cyber; DEC VAX; IBM; IBM PC XT; Prime **Operating Systems:** CMS; MS-DOS; MVS; NOS; PRIMOS: VMS; XA **Memory Required:** Minimum 640K **Distribution Media:** Magnetic tape, Diskette; Cassette **Supported Peripherals:** Printer
VENDOR AND CONTACT INFORMATION	**Vendor:** AAA Technology and Specialties Co. Engineering Software Div. 3000 Rodgerdale Road Houston, TX 77042 USA **Contact:** A.H. (Tony) Hill, Marketing Rep. Bill Taylor, Marketing Rep. **Phone:** (713) 789-6200 **Telex:** 910-881-2425 AAATECH HOU **Also Available From:** **Vendor:** COSMIC NASA Computer Software Management 112 Barrow Hall Univ. of Georgia Athens, GA 30602 USA **Phone:** (404) 542-3265

(Also listed: Documentation; Price/Terms; References and Reviews)

Print Directories of Databases and Software

Engineering & industrial software directory: a guide to computer programs in the applied sciences. New York: Engineering Information; 1988.

A guide to hundreds of programs for use by those working in engineering and/or the applied sciences. Each program is described in terms of original developer, date created, operating system required, price, hard-

---

Engineering & Industrial Software Directory, 1988. Reprinted with permission of Engineering Information, Inc.

ware requirements, documentation available, and a summary of its outstanding features.

Index and abstract directory; an international guide to services and serials coverage. Birmingham, AL: Ebsco Publishers; 1989. 2,177 p. Updated irregularly.
Covers more than 700 online databases and 30,000 periodicals. Data for periodicals include an indication of where they are indexed, while descriptions for the databases list which journals are covered in the files. There are indexes by subject, title, and ISSN.

U.S. Computer Products Center. *Directory of computer software, 1988.* Springfield, VA: National Technical Information Service; 1988; PB 88-190962/BBZ. 132 p.
Consists of descriptions of more than 1,700 programs arranged under a score of subject groups, such as chemistry, hydrology, biology, and mathematics. Each program is discussed in detail, along with information on the programming language, operating system, and hardware requirements. There are indexes by subject, by agency name, and by hardware/programming language requirements.

REFERENCE

1. Paskoff, Beth. Microcomputer software in library collections. *Library Trends.* 37(3): 315–320; 1989 Winter.
Covers many aspects of software, including selection, storage, circulation, and copyright. Gives nine criteria for software selection and sources of current reviews.

Chapter 16
Dictionaries

HIGHLIGHTS OF DICTIONARIES

Description: Sci-tech dictionaries serve primarily to provide definitions for terms; they rarely go beyond that function. Unlike general purpose dictionaries, sci-tech dictionaries usually do not provide information as to origin of a term, part of speech, synonyms, and pronunciation. They vary greatly in size and subject matter covered and are closely related to thesauri. Both English-language and foreign-language dictionaries are important in sci-tech collections.

Significance: Dictionaries are particularly indispensable parts of a science and technology reference collection. The main reason for this is the relative complexity of sci-tech terms, compared to subject fields in which the layperson might be comfortable.

Quantity: There are dozens of sci-tech dictionaries in print, including several which aim at covering all areas of science and technology. However, there is uneven coverage of certain fields or disciplines. In some subject fields there are dictionaries devoted to each of those disciplines; other fields, just as important, might be covered in very scanty fashion by the array of sci-tech dictionaries available.

Physical Characteristics: Dictionaries are usually arranged in simple alphabetical order, sometimes enhanced by drawings or by special features, such as colored plates or a few maps. Besides illustrations, about the only other distinctive feature is the numerous indexes in multilingual dictionaries, each restricted to a given foreign language.

Availability: Practically all sci-tech libraries have a few dictionaries covering sci-tech subjects, but the coverage is probably rather skimpy in smaller libraries. Certain costly sci-tech dictionaries, however, are often found only in the larger libraries devoted to these disciplines.

Retrieval: There are no particular problems involved in retrieving dictionaries. Some indexes or catalogs list "dictionaries" as a main heading while others use the term as a sub-heading (such as "Welding—dictionaries").

Intended Audience: Science dictionaries are addressed to a wide range of audiences. A few are aimed at grade school or high school students; at the other extreme are those designed for use by practicing engineers or scientists. Quite a few, however, are designed for the layperson.

Scope: A few large sci-tech dictionaries aim at listing terms in all areas of science and engineering, resulting in books with over 100,000 terms. On the other hand, some small dictionaries list only a few hundred terms that are very specific in their scope, such as a work restricted to radar or optics.

NATURE OF DICTIONARIES

The topics discussed in most sci-tech publications might be absolutely meaningless to a reader if certain basic terms were not understood. All disciplines have their own array of little-known terms. However, the number of unusual terms, as well as the frequent complexity of some of those terms, make science and technology very dependent upon clearly defined words and phrases.

Some sci-tech dictionaries cover only a small portion of the words that might need defining, while others are more comprehensive, including tens of thousands of words. A good dictionary on a topic such as computers, for example, might consist of fewer than 5,000 words and still be considered adequate in scope. An example of a dictionary with a limited scope is *Jones Dictionary of Cable Television Terminology*. Its 1,600 terms classify it as a rather small work, when compared to dictionaries that attempt to cover all areas of science and technology and might well contain over 100,000 terms. One of the best examples of a dictionary of this size is *McGraw-Hill Dictionary of Scientific and Technical Terms*. It covers more than 100 disciplines and identifies each term as to the basic discipline it involves. A work with about half this number of terms is *Chambers Science and Technology Dictionary*. It would be quite satisfactory for smaller sci-tech libraries needing a work that covered all their disciplines. Smaller numbers of terms are generally adequate for a general purpose sci-tech dictionary designed for the layperson who would be satisfied with having only the most important words listed. The determining factor in size of a dictionary is the goal set by the compilers.

General dictionaries, particularly unabridged types, often provide many bits of information besides the spelling and meaning of terms. They might include date of origin, discipline involved, synonyms, antonyms, part of speech, examples of usage, pronunciation, and other features. Sci-tech dictionaries, on the other hand, rarely give more than the correct spelling, the meaning, and perhaps the discipline involved, such as astronomy or chemistry. As will be seen later, some of them do not even give the meaning. The high cost of producing sci-tech dictionaries often limits the amount of information provided.

Although sci-tech dictionaries rarely give more than the meaning of terms, they vary a great deal in regard to how thoroughly terms are defined. Entries can be as brief as one line or as long as one column. Sometimes it is difficult to decide if a given work should be called a dictionary or an encyclopedia, because of the long and detailed definitions provided.

The usual problem of reference books, that of being up-to-date, plagues any dictionary, particularly in fast developing fields like science and technology. Unlike encyclopedias, no attempt is made to issue annual supplements to published dictionaries. Frequently the meanings of very new terms cannot be found in dictionaries; one solution is to search subject headings used in abstracting and indexing services which cover appropriate periodicals in given fields. Quite often the editors and indexers involved in preparing such services come to grips with new terms before the words are in common usage. They must make quick decisions about which terms to use. By checking the literature indexed under a new term selected by the indexers, it is possible for the searcher to find an article in which the term is explained adequately. This is not to say that indexers always pick the word that scientists and engineers ultimately choose; sometimes indexers pick a term that for some reason never becomes accepted.

SPECIAL TYPES OF DICTIONARIES

Bilingual and multilingual dictionaries are not uncommon for sci-tech subjects, although they are obviously not nearly as numerous as dictionaries (in the United States) entirely in English. For obvious reasons most bilingual and multilingual dictionaries used in this country include English as one of the languages. However, due to the widespread use of English in sci-tech publications published outside this country, it is likely that such dictionaries published abroad would also include English.

Multilingual dictionaries usually have an index for each language represented, with the entries referring the reader to the English-language equivalent of the foreign term. In such cases the most common style is to have the English-language terms listed first, followed by all the foreign equivalents on the same line. Very often the definition is omitted, so the reader would have to seek another dictionary if anything more than equivalent terms were desired. This style is used because it saves money; furthermore, it is relatively easy for readers to find the definition in a conventional dictionary should that be necessary. An example of a typical multilingual dictionary is *Elsevier's Dictionary of Microelectronics in Five Languages: English, German, French, Spanish and Japanese.* Many different languages can appear in multilingual sci-tech dictionaries.

Most bilingual dictionaries have two parts; one part is arranged by English terms, with their foreign equivalents. The other part contains a list arranged by foreign terms, with the English equivalents. Because many Americans do not have extensive linguistic skills, a foreign-language sci-tech dictionary without the English-language equivalents would not be very useful for the average library user in most U.S. libraries. Many lesser-known languages are difficult to find in any kind of a dictionary, especially bilingual dictionaries that cover sci-tech subjects.

Another type of dictionary that is useful in sci-tech libraries consists solely of abbreviations and acronyms, of which there are many for sci-tech disciplines. Many words, like "radar" or "laser," are examples of acronyms that have become accepted as words, but most acronyms require explanations of what the letters represent. Still another special type of dictionary lists trade names so that readers can learn what company produced a commercial product having a particular trade name. There are literally tens of thousands of these names.

The publishing of updated editions of dictionaries has been facilitated in recent years by the practice of some publishers of putting all the entries in computerized format; new terms can be easily inserted in the proper places and changes made on older terms as needed. The work of editing and sorting can thus be done on a daily basis, making it much quicker now to prepare a new edition. Some publishers code each entry with the name of the discipline to which a term applied. Such coding not only makes the entries more useful to readers but also enables the publisher to print only the terms coded with a particular field so as to produce a dictionary devoted just to the coded field. A database that included terms covering all of science and technology could thus be used to produce a new dictionary on a smaller scope, such as only for mathematics, botany, or civil engineering.

TYPICAL EXAMPLES OF DICTIONARIES

Chambers science and technology dictionary. 4th ed. Edited by Peter M. B. Walker. New York: Cambridge University Press; 1988. 1,008 p.
Contains some 45,000 terms in 100 fields. This modestly priced reference tool would serve small sci-tech libraries admirably. The useful appendixes include names of animal kingdoms, conversion units, and the periodic table of the elements.

Elsevier's dictionary of microelectronics in five languages: English, German, French, Spanish, and Japanese. Compiled by P. Nagy and G. Tarjan. New York: Elsevier; 1988. 944 p.
Gives the English equivalent of over 8,500 terms; there is a subject index in English plus an index for each foreign language covered. Subjects included are semiconductors, microelectronic technology, and integrated circuits.

Jones dictionary of cable television terminology: including related computer and satellite definitions. 3d ed. Compiled by Glenn R. Jones. Englewood, CO: Jones 21st Century; 1988. 108 p.
Consists of more than 1,600 definitions, all related to cable television. Some terms could be found in electronics dictionaries, while others are more obscure.

McGraw-Hill dictionary of scientific and technical terms. 4th ed. Edited by Sybil P. Parker. New York: McGraw-Hill; 1988. 2,088 p.
A long-established sci-tech dictionary, for many years recognized as one of the best. Contains over 100,000 terms, of which some 7,500 are new to this edition. Covers more than 100 disciplines. This edition features a pronunciation system. Each term is identified as to discipline to which it pertains, such as electrical engineering or biology. Appendixes include brief biographies and units of measurement.

Chapter 17
Directories

HIGHLIGHTS OF DIRECTORIES

Description: Directories are essentially lists of organizations, people, or products/services that enable users to locate quickly the information needed about the items listed. Directories usually provide a limited amount of information, rather than trying to be exhaustive in the quantity of data provided.

Significance: Directories serve a vital purpose, allowing one to locate data in a speedy fashion. The arrangement and design of directories are geared for quick retrieval of information. Directories must be updated frequently in order to be useful, with supplements or completely new editions issued according to schedules chosen by the publishers.

Quantity: Hundreds of directories are available, ranging from those covering several disciplines to those concentrating on a narrow field.

Physical Characteristics: Most directories are arranged in alphabetical order, either by the names of individuals, or organizations, or products/services. Some may have additional finding tools, such as a separate geographical index. They may be in paper or hardcover formats; supplements are usually in paperback form. Although most are in print form, several are available in either online or CD-ROM formats.

Availability: Directories constitute an important segment of reference collections in most libraries, where they are kept up to date by means of new editions or whatever supplements are available. Although some are quite expensive, others are modestly priced, thus encouraging individual purchases.

Retrieval: Directories may usually be found in catalogs or indexes under the term "Directories"; sometimes that term is used as a sub-heading under the specific type of directory sought, such as "Steel mills—Directories." Most directories are shelved in reference sections, which facilitates browsing.

Intended Audience: People of almost any background can use the average directory, although some directories are much more likely than others to interest a wide audience. Most people find them relatively simple to use.

Scope: Some directories have a wide scope, such as listing all manufacturing companies in the United States, while others might list only those manufacturers who deal in electronic products or in certain kinds of chemicals. The amount of information is traditionally brief; it is common to list names, addresses, telephone numbers, key officials, purposes of the organizations, number of members (for membership groups), names of publications (if any), amount of capitalization (for corporations), and any other type of information particularly useful to the users of the directory.

NATURE OF DIRECTORIES

The average person has used directories of one sort or another since childhood, such as telephone books and store catalogs. Directories devoted to sci-tech subjects are virtually identical to general directories in regard to method of arrangement, ease of use, and variety of available titles. Sci-tech directories also have all of the same problems as their general public counterparts, particularly the difficulty of remaining current.

There are at least three basic types of directories, whether sci-tech in nature or not. One of the most popular types identifies organizations. A directory of organizations concerned with planning tours and vacations is remarkably similar to a directory of companies having research laboratories. One is for the general public, while the other is aimed at scientists, vendors of laboratory equipment, and recent graduates of sci-tech college courses, to name a few users. In both cases it is likely the source could be searched by the organization name, perhaps by the geographic region in which located, and probably by specialty (which in the case of laboratories would be their areas of research). One difference would be the likely inclusion in the sci-tech directory of the makeup of the laboratory staff, such as the number of chemists or engineers on the staff. On the other hand, the directory of tour services might have a closely related feature, a listing of the special kinds of tours each company provided.

A noteworthy example of a directory of sci-tech organizations is the *Directory of American Research and Technology*. This annual publication lists more than 11,000 organizations related to scientific and technical research; it provides the usual directory-type information but also includes such special data as the number of researchers with doctorates and the types of research conducted. *European Sources of Scientific and Technical Information* provides data on sci-tech organizations in Europe. Now in its 7th edition, this book devotes its 25 chapters to describing groups devoted to such topics as health, engineering, and physical sciences. A detailed subject index allows for searching by fields of study featured at each installation.

Another common type of directory deals with the names of people, chosen for any number of reasons: because they were famous, or owned a telephone, or attended a certain college, or belonged to a particular organization. Again, the general directory and the sci-tech directory share much in common, with retrieval possible not only by name but also by geographic region or possibly by occupation or specialty. Some directories of this sort are very restricted as to who will be listed, while inclusion in others is relatively easy. A directory of past Nobel laureates in physics is a very discrete list, but a listing of the tens of thousands of members of a scientific or engineering society is not at all exclusive. Directories of names fall quickly out of date. Most of them are revised on an annual basis, and some have quarterly revisions. An example of a directory of this sort is the *Directory of Medical Specialists*, which is restricted to physicians who have been certified in particular health science specialties. The American Medical Association provides the means for determining the qualifications of those listed. (See Chapter 14, "Biographical Information," for more examples of directories listing the names of scientists and engineers.)

The third type of directory lists products, particularly manufactured goods, which can be as down-to-earth as chairs and tables or as complicated as expensive devices for scientific laboratories. While users of most sci-tech libraries have occasional need for the common products listed in general purpose directories, they have much greater need for specialized directories. For example, the *Thomas' Register of American Manufacturers* lists thousands of products, from nails to steam boilers. The *Register*, which is fully described in Chapter 25, "Manufacturers' Literature," is now available both online and as a CD-ROM product.

Some product directories are very specialized, listing the locations of manufacturers of particular products or the names of companies themselves. An example is *Pharmaceutical Manufacturers of the United States*. As its title implies, this directory is limited to describing the products of the manufacturers of drugs and other health care products.

The development of online databases has provided a means for producers of directories to keep their publications as up-to-date as they wish. The question of printed supplements becomes meaningless when databases can be updated on a daily or even an hourly basis. Libraries able to do online searching can profit from the new and revised data available in the online sources. Time will tell what effect the online versions will have on sales of the printed sources.

The appearance of CD-ROMs has likewise enabled directory publishers to create products that can be updated as frequently as the market will support and that can be searched in many ways not possible with printed directories. Purchasers are invariably libraries rather than individuals, particularly libraries with a strong interest in the material indexed.

TYPICAL EXAMPLES OF DIRECTORIES

Directory of American Research and Technology. New York: Bowker; 1988– . Annual.
Lists more than 11,000 organizations committed to commercial research and development, including corporations, independent laboratories, and nonprofit organizations. Each entry includes information about location, names of key personnel, size of research staff, number of researchers with doctorates, description of areas of specialization, and types of research services offered (such as government contracts, industry contracts, or consulting). Covers over 1,500 fields of scientific research. It is also available online.

Directory of Medical Specialists. Chicago: Marquis; 1939– . Annual.
This directory lists physicians who have been certified by specialty boards sponsored by the American Medical Association and other national societies. Compiled from questionnaires, the directory includes brief biographies, experience, education, internship, clinical training, and professional memberships. Arranged by state and city under the specialty board. An index is arranged by name, city, state, and specialty.

European sources of scientific and technical information. 7th ed. Edited by Anthony P. Harvey. London: Longmans; 1987. 356 p. Distributed by Gale Research Inc.
Provides descriptions of organizations in Europe that are sources of sci-tech information. The directory is divided into 25 chapters, dealing with such topics as health, metallurgy, safety, or the earth sciences. Some organizations cover a wide range of disciplines, while others are limited in scope. Along with the usual information about the location and size of the organizations listed, the directory provides the names of persons to contact, types of services rendered, and specialized fields of activity. Organizations include universities, patent offices, and information centers. The directory is indexed by subject and organization.

Pharmaceutical manufacturers of the United States. 4th ed. Edited by D. J. DeRenzo. Park Ridge, NJ: Noyes Data Corporation; 1987. 300 p.
Describes the products of more than 500 of the "leading" companies in the United States which manufacture prescription drugs, biotechnology products, and selected health care products. Data provided include the usual information about locations and number of employees, along with types of products, brand names, sales statistics, and names of subsidiary companies.

Chapter 18
Encyclopedias

HIGHLIGHTS OF ENCYCLOPEDIAS

Description: Encyclopedias for science and technology usually have the same appearance and function as the familiar general encyclopedias—they provide access to information that is arranged alphabetically by topic. They differ in that some of them are much more specific in their scope than the general types. A few of them are updated once a year by means of a separate yearbook.

Significance: Sci-tech encyclopedias, like their general counterparts, allow for quick retrieval of information because of their simple arrangement. The user may need additional sources of information, but encyclopedias are best suited to persons needing little more than definitions and a brief account of a topic.

Quantity: There are dozens of encyclopedias devoted to science and technology. Some cover all sci-tech disciplines, while others are limited to certain subject areas, such as biology, psychology, or astronomy.

Physical Characteristics: Encyclopedias devoted to science and technology have the features one expects to find in general encyclopedias. They can be single volumes or come in multi-volume sets; they consist mostly of text but may have a lot of graphic material, such as graphs, photographs, and drawings. The quality of their paper and binding is usually adequate for years of service.

Availability: Practically every library has one or more encyclopedias, but smaller libraries are not apt to have those works devoted just to science and technology, particularly the more expensive titles.

Retrieval: Encyclopedias are usually one of the easiest reference tools to use, most having a simple alphabetical arrangement of topics. The larger ones often have an overall index to facilitate locating subjects spread over many volumes.

Intended Audience: Most sci-tech encyclopedias are used primarily by people who lack an adequate background in these disciplines, particularly those who wish to find the information relatively quickly. Occasionally a person with a strong scientific background may use an encyclopedia when going outside his or her normal field of interest. On the other hand certain specialized encyclopedias are too difficult for the layperson to use; such works are aimed at a person having some education, experience, or expertise in the fields covered.

Scope: Sci-tech encyclopedias, as mentioned above, may be restricted to one or two fields, such as chemistry or astronomy. On the other hand, some attempt to cover all areas of science and technology. The narrower the scope of an encyclopedia, the more detailed it can be in its coverage of a given topic. Encyclopedias devoted to a single discipline, such as computers, can devote several columns to a topic, but lack of space would make such an extensive coverage impossible in a general sci-tech encyclopedia. In the latter case, the average entry may be no more than a dozen or so lines in length. Many encyclopedias exclude biographical material, restricting themselves to sci-tech subjects.

NATURE OF ENCYCLOPEDIAS

General encyclopedias need no description since they are so commonly used by people of all ages. However, the characteristics of encyclopedias devoted to science and technology might not be so well known. Not surprisingly, they vary a great deal in scope. Some sci-tech encyclopedias are limited to one or two general topics (such as plastics or computers), some cover broad disciplines (such as chemistry), and a few aim at covering all of science and technology. One of the best examples of a work covering all sci-tech disciplines is the *McGraw-Hill Encyclopedia of Science and Technology*, a well-written, attractively printed, and reliable encyclopedia that provides a separate volume for its detailed index of all volumes.

A number of sci-tech encyclopedias are aimed at college students majoring in science or workers in a particular area of science and technology; such a volume is the *Encyclopedia of Chemical Technology*, which is too complex to be of much help to those without at least some collegiate training in chemical engineering. Many of its articles are quite lengthy; there is a separate index volume for the set.

The scope of many encyclopedias falls somewhere in between total coverage and restriction to one or two disciplines. Whatever the breadth of their coverage of topics, most sci-tech encyclopedias provide readers with a source of well-edited information and a format that can be used quickly. These two features—reliable data and ease of use—ensure a place for encyclopedias in reference collections in sci-tech libraries and occasionally in private collections as well.

Since encyclopedias are secondary sources, they make no claim for presenting research data, material never before published. Such information is left to primary sources. Most sci-tech encyclopedias claim to be sources of information designed in general for people unacquainted with the topics searched, rather than for skilled practitioners in the field. As an aid to those wanting more information on a topic, longer articles frequently conclude with a brief list of additional readings.

Keeping encyclopedias up-to-date is a problem, particularly in fields as volatile as science and technology. Most readers are fairly tolerant of the likelihood that encyclopedias are usually not current in their treatment of fast-moving topics. Readers generally consult them in search of basic information that is apt to remain useful in spite of the years. If more recent information is needed, most readers know where to look for it. However, more and more encyclopedias are being offered in CD-ROM format, thus making it more feasible to issue supplements and revisions at periodic intervals.

In an effort to improve the currency of data, a few encyclopedias publish annual supplements, often called yearbooks. Such supplements are useful, but they are no substitute for encyclopedias that have been thoroughly and completely updated in a new edition. It would be a rare encyclopedia, however, that published a new edition before at least five years had passed, so yearbooks are normally about the only means for providing updated material. They can highlight new developments occurring since the last edition of the encyclopedia was published or since the previous yearbook was issued. Yearbooks are discussed in more detail in Chapter 12, "Annuals and Yearbooks."

Most encyclopedias are alphabetically arranged by topic, often including "See" and "See also" references to guide readers to the appropriate pages. Occasionally the topics of some encyclopedias are arranged according to some classification system devised for the purpose, but such arrangements are seldom popular with readers, most of whom are conditioned to a straight alphabetical approach.

New encyclopedias are appearing on the market at frequent intervals, and a large number of well-known works are still available in revised editions. Small libraries or those with very limited funds for sci-tech encyclopedias rely on several single-volume encyclopedias which aim at covering all of science and technology, are modest in price, and are updated periodically.

The depth of coverage of topics in these tools varies greatly from one encyclopedia to another. Many encyclopedias consist of articles less than one column in length, some articles being as short as one paragraph. Other works might devote dozens of pages to some topic that warrants such extensive coverage. Readers must seek more specialized encyclopedias if the coverage for a desired topic is too brief.

One important feature of sci-tech encyclopedias is their attention to graphics, such as drawings, photographs, charts, and tables. Use of colors is particularly effective in some instances. Because the subject matter in some sci-tech encyclopedias is usually difficult for laypersons to grasp, it is especially vital that such works use graphics whenever needed to clarify topics. Attention to this feature is the mark of a well-planned sci-tech encyclopedia.

Like most reference books intended for several years of use, encyclopedias normally have high-quality paper and sturdy bindings. While paperback versions are sometimes available, only hardbound versions should be purchased by libraries because of the length of time encyclopedias are kept before replacement and because of the likelihood of extensive use.

The quality of an encyclopedia is dependent upon the skill of the editors and contributors. Good encyclopedias include the initials of contributors, at least for longer articles. The affiliation or standing of each contributor is usually listed somewhere in the work.

A high quality encyclopedia on a given discipline is the backbone of a reference collection, and the fact that a good encyclopedia can be useful over a period of several years helps to amortize its cost over those years.

TYPICAL EXAMPLES OF ENCYCLOPEDIAS

Encyclopedia of chemical technology. 3d ed. Edited by R. E. Kirk and D.R. Othmer. New York: Wiley; 1978–1984. 24 vols.
Valuable to those who are knowledgeable of at least the principles of chemical engineering. It is considered reliable and thorough in its coverage. Articles on some major topics are occasionally more than 50 pages long; they often include extensive bibliographies. A separate volume contains a detailed index to the set.

McGraw-Hill encyclopedia of science and technology. 6th ed. New York: McGraw-Hill; 1987. 20 vols.
Probably the best encyclopedia covering all aspects of science and technology; it is aimed at the nonspecialist. It is carefully prepared, with reliable information accompanied by a large number of drawings, photographs, and charts. Contains over 7,000 articles; a separate index volume, consisting of 150,000 entries, facilitates location of information. A one-volume condensation of the multi-volume set is available; libraries not able to afford the cost of the set should find the price of this edition attractive. McGraw-Hill has also published a number of encyclopedias, each a single volume, devoted to one discipline, such as engineering or the earth sciences. The publishers make it clear that a high percentage of the contents of the single volumes are reprinted from the multi-volume set, although some new material is also included.

Chapter 19
Field Guides

HIGHLIGHTS OF FIELD GUIDES

Description: Field guides are publications that provide a visual approach to the study of the natural world. They are designed to enable the user to identify any aspect of or sample of nature covered in the particular field guide.

Significance: The study of nature is greatly enhanced if an individual is able to see actual examples of certain kinds of animals, vegetables, minerals, or other phenomena. Field guides make this possible by providing detailed information through written descriptions and through photographs or drawings.

Quantity: There are probably several thousand field guides to the United States alone.

Physical Characteristics: A typical field guide consists of a listing of creatures or features for a particular discipline, accompanied by detailed descriptions and depictions. They are generally published in a size that would readily fit into a large pocket, so that they are easily transportable. The contents are usually arranged in some logical manner, enabling information to be quickly and easily found.

Availability: Many of the field guides for popular scientific disciplines can be found in bookstores and in most types of libraries. Field guides for more advanced or rigorous disciplines can often be found in special or academic library collections.

Retrieval: Most field guides can be retrieved by the same methods used for finding monographs.

Intended Audience: There are many levels of field guides, from those written for the student or casual hobbyist to those written for professionals and researchers.

Scope: The typical field guide covers a specific and limited subject area. Subjects covered include birds, clouds and atmosphere, rocks and minerals, reptiles, plants, stars and planets, and any other part of nature that can be visually identified.

NATURE OF FIELD GUIDES

Unlike many areas of science, nature studies cannot be adequately covered in a classroom or laboratory setting. There is no substitute for studying nature in a natural setting. No one geographical area has all the examples of animals, plants, or other specimens one could want; what does exist is often not easily accessed by simple trips along major highways. In many cases such studies involve knowing about obscure locations, usually far from the beaten path. ›

In order to facilitate the study of nature, many professional associations and experts have developed guides for use in various areas of the country. Most field guides have introductions to the field covered, and include glossaries of terms, diagrams of the structural features covered, and maps of geographic distributions. These guides are written for a variety of individuals, from the novice or the hobbyist to the expert who might need some additional information. Some of the most commonly known field guides can be found in the Peterson Field Guide Series. These guides are designed for the layperson; there are dozens of volumes covering such subjects as sea life, insects, and reptiles. A competing series, licensed by the National Audubon Society, is published by Knopf.

For the expert, the field guides generally cover very specific subjects, and descriptions are highly technical. Subjects could be as narrow as a guide to trilobites for the paleontologist or a guide to a particular burial site for the paleopathologist. Physical format may also differ from the more common field guides; the book may be larger in size and designed for extended use. Also, there are usually no glossaries in guides designed for experts.

In the earth sciences, field-trip guidebooks are produced for a specific field trip. They range from informal descriptions of a particular geologic feature to very detailed articles on the area. Most include instructions on how to locate a particular spot, including mileage data. In some cases they may constitute the best description of a certain region, since nothing else may have ever been written about it. The primary audience for many of the volumes is the professional geologist, who can afford to take a trip to investigate an area of special interest.

The *Union List of Geologic Field Trip Guidebooks of North America* lists more than 6,500 field guides and is a great help in the location of a particular volume. These guides are rarely produced in large quantities, and their source may not be easy to ascertain a few years after publication; the *Union List* identifies what has been published and where to find copies. In recent years the RLIN database has begun to include such volumes.

TYPICAL EXAMPLE OF A SOURCE FOR FIELD GUIDES

Union list of geologic field trip guidebooks of North America. 5th ed. Alexandria, VA: American Geological Institute; 1989. 223 p.

This union list, compiled and edited by the Guidebooks Committee of the Geoscience Information Society, includes more than 6,500 guidebooks, published from 1891 through 1985. Entries are arranged by sponsoring organization with a chronological listing of guidebooks issued at its meetings and a list of libraries that own the guidebooks. Geographic and stratigraphic indexes are included.

TYPICAL EXAMPLES OF FIELD GUIDE SERIES

The Audubon Society field guide to . . .

A series of field guides for many of the same subjects covered in the Peterson's series is published by Alfred A. Knopf. This series provides color photographs, rather than detailed drawings, and is arranged in a different manner from Peterson's. The name, "Audubon Society," is under license from the National Audubon Society.

The Peterson field guide series.

This series, published by Houghton Mifflin Company and sponsored by the National Audubon Society and the National Wildlife Federation, covers a wide variety of topics in nature. Early volumes of the series are updated when new information is available, and all volumes have a large number of detailed drawings. Each volume has a series number. At present, topics include these series numbers: birds, including birds of Mexico, Great Britain, Europe, various parts of the United States, and bird sounds (1, 2, 8, 13, 20, 35); seashores and sea life, including shells, coral reefs, and fishes (3, 6, 24, 27, 28, 32, 36); plants and trees (10, 11, 14, 17, 22, 23, 31, 34, 37); insects (4, 19, 29, 30); rocks and minerals (7); stars and planets (15); atmosphere (26); mammals (5); reptiles and amphibians (12, 16); and identification of birds' nests (21, 25), animal tracks (9), and bird sounds (38).

Chapter 20
Government Publications

HIGHLIGHTS OF GOVERNMENT PUBLICATIONS

Description: Government publications are those prepared by a government agency or published for the government at its expense. They include printed texts, computerized records, maps, photographs, motion pictures, tables, and all other formats.

Significance: In many cases the information contained in government publications is unique, being the only source for certain data. Their subject matter is as broad as the widespread interests of government, which include all areas of science and technology.

Quantity: There are millions of government publications in existence, with tens of thousands of new ones created annually.

Physical Characteristics: Government publications have the same appearance as other publications in the same format, but the majority of them are softbound documents.

Availability: Many government publications are available from government agencies established for the purpose of promoting their sale or distribution. Some are available at no charge; others are available for examination at various types of libraries that have been named as official depositories. In addition, there are many publications with limited availability, due to security regulations.

Retrieval: Many government agencies issue their own indexes and catalogs. There are also government-prepared indexes which cover documents from several government agencies. Still other documents are indexed by commercial indexing and abstracting services. The smaller the unit of government (such as a city or county), the less likely that its documents will be indexed in any readily available publication.

Intended Audience: Government publications collectively appeal to all sectors of society and to many levels of interest.

Scope: The scope of government publications can be very narrow, very broad, or anywhere in between, depending upon the purpose for which created.

NATURE OF GOVERNMENT PUBLICATIONS

Millions of government publications cover every subject imaginable, including all areas of science and technology. They appeal to people in all walks of life, ranging from children to highly regarded scientists and engineers. Some of them are available at no cost, while others can be readily purchased. Still others may be tightly restricted as to availability because of security regulations. Government documents are published in all known formats, including graphic, computerized, and traditional printed forms. However, it is likely that the majority of them, such as patents or technical reports, are issued as softbound documents.

A large and important class of publishers is known as "government presses." This class includes international publishers, such as the World Health Organization, national government presses in most countries of the world, and regional and local governmental publishers.

In the United States the federal government is by far the most prolific source of government publications, although the state and local output is significant. The most active federal agency for the sale or distribution of documents is the Government Printing Office (GPO) in Washington, particularly its section headed by the Superintendent of Documents. The GPO handles both congressional documents as well as large quantities of publications issued by executive departments, such as the Bureau of Mines and the Geological Survey. Its chief index is the *Monthly Catalog of Government Publications.* Although it covers many fields of interest, the *Catalog*'s main value is its coverage of congressional documents and of important series issued by certain agencies. Congressional documents include Senate and House reports, hearings, bills, and laws. The range of subjects they cover is large, often including scientific and technical matters.

Other government agencies rely on the National Technical Information Service (NTIS) for the sale of their documents. Most of their publications are numbered with the prefix of PB, standing for Publication Board, a post-war agency established to promote the availability and use of technical reports. NTIS publishes a large index devoted to sci-tech subjects, known as *Government Reports Announcement and Index (GRAI).* Entries include annotations and indexes by authors, titles, subjects, and document numbers.

The huge outpouring of documents from the Department of Defense (DOD) is covered partly by *GRAI* as well as by the *Technical Abstract Bulletin. TAB,* as it is known, consists entirely of documents from the DOD, most of them in the series bearing the prefix AD (the initials standing for ASTIA Documents—ASTIA [an acronym for the Armed Services Technical Information Agency] was an earlier agency established for distributing DOD documents).

The National Aeronautics and Space Administration (NASA) has its own indexing service, *Scientific and Technical Aerospace Reports (STAR)*, which concentrates on NASA documents. Most of them bear the prefix N (for NASA) which precedes the assigned document numbers.

There are many other indexes and lists which announce and/or analyze the multitude of federal documents. In addition, the largest indexes, including those described above, have also been available as online databases for several years, further increasing the ease of retrieving documents.

Many government documents of a sci-tech nature are technical reports (see Chapter 10, "Technical Reports" for a discussion of the nature of such reports and important reference tools for use with them). While technical reports represent a very large portion of the huge number of government publications in existence, one cannot overlook the importance of other formats, such as patents, statistical surveys, maps, and photographs. Congressional documents also are important; the federal government holds many public hearings on major problems, writes laws that affect how business is conducted, and determines standards that affect how products or processes must be prepared, to name only a few examples of the scope of government documents.

There are over 1,000 depository libraries in this country that have been established to provide access without cost to those seeking federal documents. Each library determines the scope and types of documents collected. Some depositories are at universities, some in public libraries, and a few are in other types of libraries. Some libraries are depositories for technical reports or maps from the federal government.

Military security governs the availability of the many classified documents; these are accessible only to individuals with personal security clearances and a "need to know" for particular subjects. Efforts have been made in the past to establish a regular review period for the purpose of evaluating the need for continued classification of certain documents, but this has not always proved to be a viable process to eliminate over-classified documents.

TYPICAL EXAMPLES OF GOVERNMENT DOCUMENT SOURCES

Government Reports Announcement and Index. Springfield, VA: National Technical Information Service; 1946– . Semi-monthly.
Coverage consists almost entirely of sci-tech reports. Entries are arranged by a score of broad subjects, with indexes by author, title, subject, contract number, and report numbers; indexes cumulate annually. Also available online.

Monthly Catalog of United States Government Publications. Washington, DC: Government Printing Office; 1895– . Monthly.
Covers both congressional and executive department documents on a broad range of subjects, including science and technology. Arranged by issuing source, with indexes arranged by personal authors, titles, subjects, and some document numbers; indexes cumulate annually. Covers many major agencies involving sci-tech activities, such as the Geological Survey and the Bureau of Mines. Also available as an online database.

Chapter 21
Guides to the Literature

HIGHLIGHTS OF GUIDES TO THE LITERATURE

Description: Guides to the literature are essentially extensive bibliographies that list and discuss the types of literature pertaining to one or more disciplines. Such guides normally describe all the formats in which the literature is available and list outstanding examples of each format. Those dealing with science and technology are similar to those for other disciplines. A few guides give much more information than others about the characteristics of the formats of literature they cite, such as explaining the difference in purpose and use of patents compared to the nature of handbooks. Some guides barely mention such characteristics.

Significance: These guides provide an excellent way to learn what the outstanding sources of information are for the disciplines they cover. Being aware of the information sources available in a given discipline should enhance the reader's ability to locate needed information in an efficient manner.

Quantity: At any one time there are usually several dozen guides to sci-tech literature available, in addition to many more which deal at least in part with sci-tech disciplines.

Physical Characteristics: Guides usually resemble conventional bibliographies since so much of their contents consists of citations (often including annotations) of various formats of literature. They invariably contain indexes and may have either hardcover or softcover bindings.

Availability: Most sci-tech libraries are apt to have a few guides to the literature, which are reasonably priced for the most part.

Retrieval: Guides to the literature are usually listed as such in most catalogs, although guides of a more specialized nature may appear under the name of the discipline involved rather than under the more general heading.

Intended Audience: Probably the most avid users of sci-tech guides to the literature are librarians working with such collections. Next in amount of usage would come students studying any of these disciplines. Beyond the student level, science and technology guides to the literature are not heavily used, primarily because most sci-tech professionals believe themselves already aware of the important literature in their field (which belief is not always well founded).

Scope: A number of the current guides to the literature deal with all branches of science and technology, some in great detail. Others are limited to one or two disciplines, such as chemistry, mechanical engineering, or physics/mathematics. Guides also vary in their depth of coverage, some listing far more citations than their competitors. Some are restricted to English-language materials, while others are not.

NATURE OF GUIDES TO THE LITERATURE

One of the most useful formats for those desiring to learn about the information sources of any discipline is the guide to the literature. This is particularly true for science and technology; many literature guides are prepared for sci-tech disciplines. Some guides cover all phases of science and technology (as this book does), while others might be devoted to a single discipline, such as chemistry or electrical engineering.

The main purpose of the ideal guide to the literature is to help those unacquainted with the information sources in the disciplines under consideration to answer the following questions:

- What are the most important sources?
- What are their scopes and arrangements?
- What type of source is used for particular purposes?
- What are the strengths and weaknesses of certain sources?
- How are the various information sources used?

Guides to the literature make a good starting place for becoming familiar with the sources available. Not all guides to sci-tech information sources present much description of the various sources; some are primarily comprised of citations of specific works, perhaps including a brief annotation for each item. However, guides to the literature are more valuable when they do include such discussions of the nature of the sources.

Ideally, scientists and engineers not familiar with information sources in their disciplines should use guides to the literature to gain a better understanding of what sources exist, but scientists or engineers will rarely take the time to look over guides of this sort. While it is true that many sci-tech professionals are quite familiar with the tried and true sources they have used over the years, it is likely that most of them are not familiar with all the sources that are available in their field of work, particularly newer formats or less known reference works. Because of the reluctance of most scientists and engineers to use guides to the literature, the most likely user would be a science librarian or the library school student who aspires to work in sci-tech fields.

Besides differing in scope, guides to the literature usually differ in the types of information formats that they cover. For example, a guide to the literature of mathematics would probably not even mention patents, since patents have so little value to scholarly mathematicians. Guides to the literature for civil engineers might not place much emphasis on dissertations, which tend to be of greater interest to scientists dealing with basic research than to practical engineers in the field.

The styles in which guides to the literature are written are worth noting. Some have a rather chatty manner, as if some knowledgeable person were describing the information sources in a conversation with a person seeking instruction and guidance. Such books are perhaps interesting to read, but they are not designed for tutorial purposes. It is difficult to get a quick overview of a discipline when the style of a guide is so informal. Another problem is that formal citations of examples are rarely given in books using the informal format; instead, the titles may be casually referred to in a paragraph in which many other sources are discussed. Indexes for this style of literature guide are rarely complete, making it difficult to locate a particular title quickly.

The other style emphasizes formal citations for each source (providing such data as author, title, publisher, and publishing date), often including a brief abstract or annotation describing the contents of each item. Indexes for this style of guide tend to be comprehensive, making it easy for the reader to locate a particular citation. This style is much better suited for tutorial purposes than the informal type of presentation.

Still another difference is the organizational plan used for guides to the literature. One method is to arrange chapters by the disciplines covered in the book, such as a chapter for all information sources on chemistry, followed by a chapter on mathematics, and so on. Within each chapter the sources would be subdivided by format, such as dictionaries, then encyclopedias. An example of this arrangement is found in *Information Sources in Science and Technology* by C. D. Hurt. Aside from one chapter on interdisciplinary sources, each chapter is devoted to a single discipline, such as physics, then subdivided by format.

The other method is to arrange chapters by the formats of the sources (dictionaries or encyclopedias), with each chapter subdivided by disciplines (physics or electrical engineering). The latter method simplifies use of the guide in classrooms since the instructor can discuss the features of a particular format, such as handbooks, then use the examples in that chapter as illustrations of the characteristics of handbooks. An example of this style is found in *Scientific and Technical Information Resources* by Krishna Subramanyam. Chapters provide detailed information about each format, such as handbooks, then come subdivisions into various disciplines, such as engineering or botany.

Guides to the literature of science and technology first began to appear in the United States around 1920, and the earliest ones tended to deal with chemistry. This is not surprising in view of the traditional attention chemists have given to sources of information. Most guides are published as separate books, but occasionally they might appear as periodical articles or as an entry in a sci-tech encyclopedia.

TYPICAL EXAMPLES OF GUIDES TO THE LITERATURE

Hurt, C. D. *Information sources in science and technology.* Englewood, CO: Libraries Unlimited; 1988. 362 p.
Lists more than 2,000 titles, primarily reference books which collectively cover all areas of science and technology. Only English-language books are cited. One chapter covers multi-disciplinary works, while the other chapters are each devoted to a single subject area, such as physics, geology, or chemistry. Chapters are subdivided into different formats, such as tables, handbooks, or encyclopedias. There are two indexes, one for authors/titles and one for subjects. Each entry has an abstract. The nature of each format is briefly described.

Subramanyam, Krishna. *Scientific and technical information resources.* New York: Dekker; 1981. 416 p.
Provides an excellent background on the history and nature of the major sources of sci-tech information. Has an abundance of examples; the indexes are reasonably thorough, although not as complete as could be desired.

Chapter 22
Handbooks

HIGHLIGHTS OF HANDBOOKS

Description: Handbooks are collections of data in both textual and numerical format that their compilers have chosen to represent the most important information on the topics covered by the books. Handbooks are written for experienced people in the field involved and are designed as reference tools that are relatively quick to use.

Significance: Handbooks are indispensable tools for users who already understand the principles of a particular field but want a quick source for a tabular value, a formula, a basic principle, or an account of an accepted practice. They are usually confined to one discipline or a small topic, and they are generally accepted as authoritative sources.

Quantity: There are hundreds of handbooks in existence, some much more specific in their scope than others. They are more common in engineering or applied science than in the pure sciences. A few handbooks are available online.

Physical Characteristics: Handbooks usually have quite a number of tersely written chapters along with many tables, charts, and formulas. They tend to be printed on high-quality paper and have sturdy bindings because their useful life is normally rather long. Many tend to be small in size, so that they may be read in any location. The online versions tend to mirror the style and format of the printed volumes.

Availability: Handbooks are quite common in certain sci-tech libraries, particularly those which include technology, less so in libraries devoted to the pure sciences.

Retrieval: Most catalogs list handbooks as a sub-heading under the name of the topic(s) covered, as well as by titles.

Intended Audience: Handbooks are used most frequently by experienced people or experts who want a quick review of a topic or seek numerical values from tables. They are too difficult to be of use to beginners in a particular field; they are not written as tutorial aids.

Scope: A handbook rarely covers more than one discipline, such as physics or chemistry, and most do not attempt to cover as much information as that. As fields develop and become more complicated, handbooks tend to encompass fewer topics so that the material can be covered in a detailed manner. A typical handbook might be concerned only with highway engineering or noise control.

NATURE OF HANDBOOKS

In certain sci-tech disciplines the need for a compact compilation of formulas, tables, and charts resulted in the creation many years ago of a format known as handbooks. They became very popular in certain disciplines, and now hundreds of them are available. Engineering and the biosciences are two subject areas that rely heavily on handbooks.

Handbooks are written primarily for experienced people, not for the neophyte. A person wanting to learn about the basic concepts of some sci-tech topic would be ill-advised to attempt to use a handbook. Neither would such a source be suitable for searching for some new research results. Handbooks are essentially collections of what have proved to be useful data designed for professionals in a given field, particularly for experts who need a source that could be used to remind them about a certain formula, or to review the characteristics of a material involved in some project, to cite some typical uses. Quick access and reliable data are probably the two most important features of handbooks. The more theoretical a subject, the less likely there will be handbooks published in that field simply because there is virtually no demand for them.

It is rare to see a handbook that attempts to cover more than one major discipline. A handbook attempting to include all of the physical sciences, for example, would be most unusual. The largest subject area for a typical handbook might be all of physics, or all of electrical engineering. Some are devoted to even smaller subjects, such as piping or semiconductors. For example, a well-known handbook on properties of dangerous materials is aimed entirely at discussing the hazards presented by these substances and how they should be handled. Entitled *Dangerous Properties of Industrial Materials*, it is generally considered the most authoritative work on this topic.

Editors of handbooks are typically very experienced in their fields, and they select other specialists to write individual chapters. Charts, graphs, and tables are invariably used to summarize important data, while the texts are usually tersely written. Since there is no need for making a topic clear to a beginner in the field, elaborate explanations and detailed clarifications of points are not used.

A few large publishers publish the majority of sci-tech handbooks, although smaller publishers and some professional organizations also produce them. In recent years a new style of handbook is the multi-volume set, with each volume devoted to a particular phase of a topic. Such sets can be quite thorough in their treatment of a subject. The *Welding Handbook*, for example, provides a separate volume for welding equipment, techniques of welding, and similar topics. Volumes are issued and updated on an irregular basis, but the multi-volume format facilitates revisions since each volume is of a manageable size.

Updating of major handbooks usually occurs every five or six years. Because handbooks are expensive to publish, frequent editions are not very likely.

One recent version of handbooks involves incorporation of the data they contain on CD-ROMs. CD-ROMs require purchase of a special player, but they can be searched at no extra cost over the original purchase price, no matter how long a search is involved. Often sophisticated search strategies can be used to identify elusive data not easily found in the print version.

One mark of a good handbook is the citation of recent references at the end of each chapter, offering readers a source for more detailed descriptions of a topic. Still another sign of high quality handbook publishing is a thorough index. Traditionally the best handbooks have had excellent indexes, which are very important for bulky, fact-filled reference works such as these.

Some books bearing the word "handbook" in their titles really aren't handbooks of the sort described in this chapter. At best some are collections of major tables and charts, with little or no text.

TYPICAL EXAMPLES OF HANDBOOKS

Dangerous properties of industrial materials. 7th ed. Compiled by Irving Sax and Richard J. Lewis. New York: Van Nostrand Reinhold; 1989. 3 vols.
  Contains chemical data and hazardous characteristics for 20,000 substances. Other features include citation of the Chemical Abstracts Service Registry Numbers, a multilingual index of synonyms, and clinical data on human beings and laboratory animals.

Welding Handbook. 7th ed. Miami, FL: American Welding Society; 1976– .
  A multi-volume set in which each volume is devoted to a particular topic, such as welding processes, fundamentals of welding, and welding equipment. Updating continues for each volume, although on an irregular schedule.

Chapter 23
Histories and Archival Materials

HIGHLIGHTS OF HISTORIES AND ARCHIVAL MATERIALS

Description: Historical materials are publications written specifically to record important events in times past. Archival materials are usually thought of as records, correspondence, and other nonpublished materials that have been saved because of their historical value.

Significance: Historical materials not only preserve the records of important accomplishments in science and technology but also provide lessons, which, if heeded, enable us to avoid the mistakes of the past. Likewise, archival materials not only record transactions that took place in the past but also may shed light on conditions and problems of previous generations.

Quantity: Thousands of books have an historical purpose; a few journals are also devoted to this subject. As for archival materials, there are millions of records in various archives.

Physical Characteristics: Historical materials outwardly resemble other publications, while archives can take numerous formats: correspondence, reports, slides, video and computer tapes, illustrations and so on.

Availability: Historical materials are usually freely available to library users, except that there would undoubtedly be restrictions on access to very rare items. Archival materials are usually available only to certain people in an organization who have some legitimate need to see and use them.

Retrieval: Some classification schedules have a class for historical materials, usually not restricted to science and technology. In some cases a sub-heading of "—History" might be found under a given branch of science and technology. Archival materials are usually indexed in separate systems from books and journals; their specificity generally requires special subject treatment.

Intended Audience: Historical materials are written for a wide range of users, for the layperson as well as the scholar. A four-volume history of chemistry is not apt to interest the casual reader, whereas more and more scientists and engineers are writing books for the average adult. Archival materials are designed to be used by people within an organization who need to trace previous records, as well as an occasional nonaffiliated person, such as a writer who needs access to the material for gathering data for a book.

Scope: Some histories have the ambitious goal of covering all areas of science and technology, but it is much more common to find books dealing with smaller topics, such as the development of computers or the history of bridge building. Archival materials may cover a wide range of organizational activity, often in excruciating detail, such as all past payroll records or 20 years' worth of purchase orders.

NATURE OF HISTORICAL AND ARCHIVAL MATERIALS

Science and technology have been part of our heritage for many centuries, reaching back to prehistoric times. Over the years humanity has kept track of important discoveries and inventions, although there is little concrete evidence of the dates and persons involved in many early events that took place before the use of printed or graphic records made accurate histories possible. We will never know who was the first person to make a crude iron spear or who was the first to discover how to make a campfire by rubbing two sticks together so that the friction ignited some dry substance. For centuries many devices and discoveries that are now accepted as commonplace went unrecorded as to the time, place, and persons involved.

Some of the earliest scientific and technical events date back to the ancient Egyptians, perhaps as long ago as 4000 BC, when they devised a calendar that enabled them to predict the dates of the annual flooding of the Nile River. Ancient manuscripts and scrolls provide some idea of the date at which developments like this occurred, but these records are sketchy and incomplete. Information about ancient events can also be found in the clay tablets made by the Babylonians, one of the earliest written records we have of technical developments. For example, the tablets show that this civilization was skilled in arithmetic over 4,000 years ago. Nevertheless historical proofs of inventions and advances made by ancient civilizations are few in number.

It was not until the invention of printing around 1450 that records with any degree of accuracy and completeness could be kept. Since then, great quantities of data concerning the times, personages, and places of inventions and discoveries in the sciences and engineering have been kept. Several formats are commonly used; historical accounts appear in such sources as books, periodical articles, and biographies. Books and journal articles written expressly for the purpose of recording or analyzing the history of science and technology are available in large quantities. One highly regarded writer, Isaac Asimov, has written a number of books on the history of science; his work entitled *Asimov's New Guide to Science* serves as an excellent reference source for tracing the history of all branches of science over the ages.

Other writers have prepared guides to the development of scientific and technical inventions and discoveries. A book by Ellis Mount and Barbara List provides a record of 1,000 such events, with indexes for names, nationalities, and dates of the persons involved. At least one journal, *ISIS*, deals exclusively with the history of science. A few other periodicals concentrate on the development of technology. Some universities even have departments devoted to the history of such disciplines.

Closely related to materials on the history of science and technology are those devoted to the philosophy of these disciplines. Books on this subject are likely to be of interest to philosophers and historians rather than to scientists. An example is the work co-authored by Jean Piaget and Rolando Garcia, entitled *Psychogenesis and the History of Science*. It is a scholarly treatment of a complex topic.

Literature on the relationship of science and technology to society is related in many ways to historical records dealing with certain inventions and developments. For example, the impact of machines in Europe during the Industrial Revolution, beginning around 1750, has been studied carefully by economists, historians, and sociologists. Many new ways of understanding our present society have evolved from examinations of historical records of early scientific and technical developments.

It is not always easy to interest modern practitioners of science and techology in the history and philosophy of their disciplines, despite the fact that a study of such materials would give them a broader perspective of their professions. In many cases, historians, philosophers, and sociologists are the ones who care most about such topics. The general public has also shown an interest in the history of science and technology, such as curiosity about who invented certain devices or when the first model of a particular invention was created. However, the public has little or no interest in discoveries or inventions of a minor nature, particularly those beyond the comprehension of untrained people. It would be hard to expect a layperson to be interested in the date of the invention of the first centrifuge if the person had no idea what a centrifuge was.

Besides literature written for the express purpose of discussing the historical aspects of these subjects, there is an entire body of literature and records prepared by practitioners as part of their research and development efforts. As time passes, older records of this sort automatically become part of the history of the disciplines, even if not intended to be such when created. In some libraries issues of certain periodicals containing the original articles of famous scientists are so popular and so apt to be mutilated by avid but unscrupulous readers that the issues have to kept under lock and key. An example is the original papers on relativity by Einstein that appeared in certain European journals devoted to physics. Needless to say, such mutilations of the printed records of the history of science are not welcomed by librarians.

In addition to the formal publications carrying original papers, there is the informal type of record, such as diaries and correspondence of those involved. Biographies of major scientists and engineers often include an analysis of their personal writings, in many cases revealing events and attitudes previously unknown.

Another informal type of record that is becoming more important in the exploration of the history of science and technology is the oral history. Oral histories are created by interviewers who record, on audio- or video-tape, conversations with individuals who were involved with major discoveries or events. The interviewees are usually of advanced age, and their careers span a number of decades. They are often privy to occurrences, conversations, and details that are not reported anywhere in the literature. The recollections of these individuals provide insight into how discoveries were made, and what influenced these discoveries. The interviews are usually transcribed, and printed copies can be made available to researchers.

The first oral history program was established at Columbia University in 1948, and since then a number of organizations have established oral history programs to capture the remembrances of individuals, both famous and not-so-famous. There is also the Oral History Association, created in 1966, which has developed guidelines for conducting oral history programs.

Obviously the older the field, the more voluminous the history of that field. For example, a complete history of civil engineering would have to range from accounts of the construction of ancient buildings, such as the pyramids around 2700 BC, up to the present, whereas a history of computers would deal merely with events occurring chiefly since the 1940s, aside from references to a handful of eighteenth-century devices that used punched cards or in some way predated twentieth-century inventions.

Histories of science and technology play a very useful role in bridging the past with the present and giving us a better understanding of our heritage.

TYPICAL EXAMPLES OF SCI-TECH HISTORIES

Asimov, Isaac. *Asimov's new guide to science.* Rev. ed. New York: Basic Books; 1984. 940 p.
An excellent, reliable review of developments in all branches of science, ranging from ancient eras to the present time. Provides names of scientists, dates, and locations for each discovery described. Contains thorough indexes.

ISIS-International Review Devoted to the History of Science and Its Cultural Influences. Philadelphia: History of Science Society, Inc.; 1913– . Quarterly.
An outstanding journal devoted to the history of science; it also touches on the philosophy of science. Consists of papers, news of the profession, and book reviews.

Mount, Ellis; List, Barbara A. *Milestones in science and technology: the ready reference guide to discoveries, inventions, and facts.* Phoenix, AZ: Oryx Press; 1987. 141 p.
Contains brief accounts of what the compilers felt were the 1,000 most signficant inventions and discoveries in all branches of science and technology from prehistoric times up to the present. There are four indexes: the year the event took place, the nationality of the persons involved, the names of the persons, and the broad category in which the discovery belongs.

Piaget, Jean; Garcia, Rolando. *Psychogenesis and the history of science.* New York: Columbia University Press; 1989. 309 p.
Written for scholars interested in the concepts of Piaget on cognitive structures and their development. Topics covered include the historical development of geometry, algebra, and mechanics.

Chapter 24
Manuals

HIGHLIGHTS OF MANUALS

Description: Manuals provide, in concise form, explanations of procedures or outline, in detail, how a piece of equipment works.

Significance: Some manuals provide the user with invaluable information on the proper or approved methods to perform a procedure. Other types of manuals describe how equipment is supposed to operate, or how the user can, most effectively, use the equipment. Often mistaken for handbooks, many manuals are overlooked as sources of information for use in learning or reviewing procedures or processes.

Quantity: Thousands of manuals exist, many of which fall into the "how-to" books category.

Physical Characteristics: Manuals are generally produced in compact and easy-to-carry form; they are often small enough to fit into a pocket. Manuals are designed to be used in any location with ease, containing specific and detailed information along with comprehensive tables of contents and detailed indexes.

Availability: Depending on the sophistication required, most manuals are available in many types of libraries and bookstores. Very technical manuals generally are available in libraries with special, subject-oriented collections.

Retrieval: There may be some difficulty in identifying manuals, since there has been confusion in the publishing field over the terminology of "manuals" and "handbooks." Often, a manual may have the word "handbook" in the title, and vice versa. Certainly, the "how-to" works can be identified as manuals. Another name for these publications is "instruction book." For other works, the text must be examined to determine whether descriptions of processes, etc., are present.

Intended Audience: There are manuals for the inexperienced user or novice as well as for the expert. Some disciplines, such as engineering, require highly technical manuals; "how-to" books are often written for people who have little or no knowledge of the subject or process.

Scope: Manuals in science and technology generally cover a specific discipline, or a narrow aspect of a discipline. Because they cover processes and procedures, they do not fall out of date for the material they provide, although newer processes and procedures, perhaps developed through new technology, require new editions to be published.

NATURE OF MANUALS

Manuals are familiar to most people because they provide instructions on the operation of equipment or the proper method of executing a procedure. Examples of the most common types of manuals include the operating manuals provided with new television sets, or kitchen equipment, or the supermarket variety of gardening, landscaping, or home repair books.

In science and technology, however, it may be difficult to identify manuals, in many cases because they can be very similar to handbooks. Many glossaries define manuals as handbooks and handbooks as manuals; yet there is a difference. Handbooks are primarily for the expert and contain compilations of data without introductory or explanatory material. Manuals, however, generally tend to have a great deal of explanatory material, describing necessary information needed to perform some procedure or to operate some type of equipment. Manuals can be written for the novice, for the expert, or for anyone in between.

If the primary emphasis of a work is to describe how a piece of equipment operates, it is a manual and is easy to identify. Such a manual generally offers detailed descriptions of components of the equipment with diagrams of important parts. It also describes, sometimes in detail, how the equipment performs when properly used and provides suggestions on probable causes and remedies for malfunctions. Some manuals are produced by the manufacturers or designers of the equipment.

Manuals that describe proper or accepted methods of performing procedures are more difficult to identify. These can be produced by almost anyone, depending on the subject covered. An example of a well-known handbook covering procedures for health sciences personnel is the *Merck Manual*. Such manuals may be very general and created for the novice in order to introduce a topic. "How-to" books generally fall into this category. For example, books describing how to raise animals, or how to perform simple laboratory experiments, or how to create projects for science fairs may be manuals for the novice.

Manuals may be very specific and created to aid experts in using specific types of equipment. The audience for these manuals is expected to have an advanced educational or experiential background. Therefore, the explanations tend to be technical and omit basic information that the experts in the field should know. The audience for whom the manual is written is usually identified in the introductory chapters. A manual for a growth chamber, edited by Robert Langhams, is an example of a manual for a specific type of equipment. The book provides specific information on the use of growth chambers as well as data on equipment manufacturers.

Because of the wide variety of manuals available in the marketplace, and because of the wide areas of coverage in science and technology, it is difficult to make generalizations about this type of material. Care should be taken to ensure that manuals match the level of knowledge, as well as the needs, of the user. Care should also be taken to ensure that what is purchased as a manual actually is a manual and not a handbook or other type of information source.

TYPICAL EXAMPLES OF MANUALS

Langhams, Robert W., ed. *A growth chamber manual: environmental control for plants*. Ithaca, NY: Cornell University Press; 1978. 222 p.
Contains specific data on growth chambers and the equipment used to regulate precise environmental conditions within them. Written for researchers, this manual includes a list of manufacturers of growth chambers and guidelines for reporting studies in controlled environment chambers.

Merck manual of diagnosis and therapy. 15th ed. Rahway, NJ: Merck; 1987. 2,696 p.
Created for practicing physicians, medical students, interns, residents, and other health professionals, this manual provides proper procedures for diagnosis and treatment of conditions and/or diseases. It is divided into 24 broad subject sections and includes an extensive subject index. It is revised approximately every five years.

Chapter 25
Manufacturers' Literature

HIGHLIGHTS OF MANUFACTURERS' LITERATURE

Description: Manufacturers' literature is prepared primarily to increase sales of products. It is issued in many formats and is considered rather ephemeral in nature.

Significance: Despite the commercial purposes for which this literature is designed, there is value in these materials as sources of information about the nature of the manufacturers' products, including their capabilities, their construction, their limitations, and their cost. This literature often presents unique data, found in no other source. It appears in a variety of formats, but it falls out of date in a relatively short time.

Quantity: Hundreds of thousands of companies prepare and distribute their own catalogs and brochures. The needs of the users of this literature determine which publications would be appropriate to obtain.

Physical Characteristics: The amount of information found in such literature varies so much that the manufacturers may go far beyond a single sheet or two, ranging in some cases to large loose-leaf binders. Some of this material is issued on very expensive paper, usually in color, while less expensive paper and single sheets may suffice for other manufacturers. A few printed directories of manufacturers also include some pages of their catalogs or brochures, but in very limited quantities. In a few instances, companies sell subscriptions to microfiche versions of manufacturers' literature in selected fields.

Availability: Most of this literature is available at no cost, but some manufacturers feel justified in charging a fee for this type of promotional material.

Retrieval: There are few ways to retrieve a given piece of this literature. One can use directories to locate the names of manufacturers of particular products, but, aside from a few large directories that may include some pages of selected manufacturers' catalogs, one often has to get in touch with the manufacturers to obtain their catalogs. In some instances local sales representatives may be able to furnish copies upon request.

Intended Audience: Manufacturers' literature is primarily aimed at design engineers, architects, or purchasing agents—literally anyone who might buy a product. These people need specifications of particular products in order to make wise purchases or recommendations for purchases by others.

Scope: Most manufacturers' literature falls into two classes: general catalogs, which describe from dozens to hundreds of items, and advertising brochures, which feature only a few products. The latter class furnishes far more information on the product than do the multi-item catalogs. The kinds of products listed ranges from a 50-ton metal press to a tiny pair of laboratory tweezers, or even a living one-cell animal.

NATURE OF MANUFACTURERS' LITERATURE

Manufacturers' literature, often referred to as trade catalogs, is indispensable in certain types of libraries. For example, libraries (or information centers) maintained for engineering companies, or architectural firms, or research laboratories, must be able to provide their clients with current data about manufactured objects, ranging from huge pieces of equipment for building roads to tiny microchips for making computers. In many cases it is not enough to have literature from only one manufacturer of a given product, since comparisons of cost, size, operating characteristics, and reliability are necessary for an engineer, scientist, or architect to make the best selections for a company or a research project. The information must be current and quickly available.

Thus many sci-tech libraries routinely include collections of trade catalogs along with other types of literature. The catalogs are generally stored in vertical files. Periodic letters may have to be sent to obtain new catalogs, which is often an ongoing project throughout the year. As an alternative, a few publishers sell subscriptions to microfiche versions of manufacturers' literature in selected fields of interest, with updates furnished periodically. While such a service would not be the only source for a library to use, it might save a great deal of time for the librarian if most of the catalogs needed are obtained through participation in such a plan. An issue of *Science & Technology Libraries* that is devoted to the subject of trade literature includes an article by Bruce Norton that describes the services and products of Information Handling Services, a commercial firm that provides libraries with microfilm sets of trade catalogs that are tailored to fit specific needs.[1]

Trade catalogs come in a variety of sizes and appearances, and they differ in their completeness of data and overall value to readers. Some are nothing more than one or two pages, perhaps giving only sketchy information about the product. Others give detailed technical information, such as the operating characteristics, dimensions, weight, or purposes for which the items were designed. A few catalogs may be lengthy; some may be offered in loose-leaf binders, with updated supplements provided to interested readers. Quite often manufacturers' literature is the only source for certain information. Imagine the difficulty of trying to find the size and weight of a particular steam roller by perusing books and journals. Such information is unlikely to appear anywhere but in a manufacturer's catalog for that particular product.

Information about prices of materials frequently does not appear in the trade catalog itself. In order to keep such data up to date, manufacturers often print separate price lists, which can be easily and inexpensively revised, rather than having to revise more expensive catalogs. On the other hand, catalogs that appear throughout the year, such as those issued quarterly, are more apt to include prices as part of product descriptions than those catalogs published less frequently.

One note of caution about trade catalogs involves their accuracy. While there are state and federal laws regarding truth in advertising, a few unscrupulous companies might try to mislead potential customers with the wording of descriptions in the catalogs. One common complaint is the failure of catalogs to state certain important facts about a product. For example, it would be an unusual catalog that mentioned the faults of a product, such as being excessively noisy, difficult to repair, or prone to frequent breakdowns. Only the reputation of a company can give guidance as to the likelihood that the information given in the catalog is reliable and complete. Warranties issued for the product help customers who have problems with the equipment, but most people don't want to go through the annoyance of getting something replaced or repaired soon after acquiring it.

Rather than create indexes to their trade catalogs, some libraries may rely on nationally known directories of manufacturers, such as *Thomas Register of American Manufacturers*, which lists thousands of companies and provides a detailed index of sources for various products. In recent years this publication has included bound volumes of trade catalogs from those companies wishing to pay for this method of presenting their catalogs to potentional buyers. A CD-ROM version was recently introduced. A publication that is very similar in scope to *Thomas Register* is *MacRae's Blue Book*, which also includes catalogs for some of the companies it lists. Figures 25-1 and 25-2 represent typical pages from this outstanding publication.

Figure 25-1 shows how companies are listed under the names of the products they manufacture, and Figure 25-2 is a reproduction of a page describing particular products of a manufacturing company.

Figure 25-1. *MacRae's Blue Book*—Product Listings

ROBOTS 1121

APM Hexseal, 44 Honeck St., Englewood, NJ 07631-4134, (Self-sealing) (1)
Admiral Screw Co., 2240 W. Walnut, Chicago, IL 60612-2218 (1)
Albany Products Co., Inc., 150 S. New Boston Pk., Woburn, MA 01801-6204, (Alum., stainless, brass, bronze, monel) (25)
American Monarch Corp., 5900 Park Ave., Cleveland, OH 44105-4945 (1)
American Rivet Co., Inc., 11300 W. Melrose St., Franklin Park, IL 60131-1323 (6-1/2)
Armour Screw Co., 7350 W. Agatite, Chicago, IL 60656-4704
Arrow Rivet Co., Inc., 60 Columbia Terrace, Braintree, MA 02184-1347 (1)
Art Wire/Doduco, 9 Wing Dr., Cedar Knolls, NJ 07927-1006, (Contact, trimetal composite) (50)
Automatic Fastener Corp., Cosgrove Industrial Park, Branford, CT 06405, (Cold headed) (20)
Banner Screw Co., Inc., 2901 W. Montrose Ave., Chicago, IL 60618-1403 (2)
Brainard Rivet Townsend/Textron, 222 Harry St., Box 30, Girard, OH 44420-1759, (Plier, Large Solid etc.) (6-1/2)
Brunner Drilling & Manufacturing Inc., Box C, Elroy, WI 53929 (10)
CEM Co., Inc., Div. of Spirol International Corp., Box 179, Danielson, CT 06239-0179, (Long, roll formed with seam) (20)
CSM Screw Products Co., 2105 N. Southport Ave., Chicago, IL 60614-4017 (1)
Cherry Fasteners, Div. of Textron Inc., 1224 E. Warner Ave., Santa Ana, CA 92705-5414
Circon Corp., 749 Ward Dr., Santa Barbara, CA 93111-2918, (Instrument-microminiature) (13)
Clark Brother Bolt Co., Canal St., Milldale, CT 06467 (6-1/2)
Connecticut Manufacturing Co., 115 Benedict St., Waterbury, CT 06706-1002 (3-1/4)
Elco Industries Inc., 1111 Samuelson Rd., Box 7009, Rockford, IL 61109-3641, (Steel, brass and copper) (132-1/2)
Goodrich Co., B.F., Engineered Products Group, 500 S. Main St., Akron, OH 44318, (Blind) (100)
Hi-Shear Corp., 2600 Skypark Dr., Torrance, CA 90505-5314 (99)
Industrial Precision Fastener Co., 518 Hankes Ave., Aurora, IL 60505-1802 (2)
International Tool Co., The, 500 N. Smithville Rd., Box 31276, Dayton, OH 45431-1069 (20)
Intra-National Screw & Bolt Co., 6512 N. Clark St., Chicago, IL 60626-4002
Jessen Manufacturing Co., Inc., Box 1727, Elkhart, IN 46515-1727 (17)
Lincoln Manufacturing Co., Inc., 2617 W. Fletcher St., Chicago, IL 60618-7109 (2)
Micro Plastics Inc., Hwy. 178, N., Flippin, AR 72634, (Nylon with steelpin)
Mid-Continent Screw Products Co., 3701 W. Lunt Ave., Chicago, IL 60645-2615 (10)
Milford Rivet & Machine Co., 857 Bridgeport Ave., Milford, CT 06460-3139 (10)
Nelson Div. of TRW Inc., Toledo Ave. & E. 28th St., Lorain, OH 44055, (Stud) (100)
Northwest Bolt & Nut Co., 4250 Eighth Ave. N. W., Seattle, WA 98107-4503 (20)
Prairie State Screw & Bolt Corp., 3685 Woodhead Dr., Northbrook, IL 60062-1816, (To order) (4)
Shingle Belting, 500 Gravers Rd., Plymouth Meeting, PA 19462-1709, (Dryer felt) (25)
Southco Inc., 261 Brinton Lake Rd., Concordville, PA 19331 (10)
Specialty Screw Corp., 2801 Huffman Blvd., Box 5003, Rockford, IL 61103-3906 (10)
TCR Corp., 1600A 67th Ave., N., Minneapolis, MN 55430-1702 (54)
USM Rivet Div., Emhart Corp., 510 River Rd., Shelton, CT 06484-4517, (To order) (100)
Venas Tool Inc., 79 Hayes St., Torrington, CT 06790-6816, (Also screws, bolts, fasteners)

Rivets, Split: *See Rivets, Clinch, Bifurcated; also Rivets, Pronged; also see Rivets, Slotted, Clinch*

RIVETS, Spot or Decorative

Standard Rivet Co., 71 A St., S. Boston, MA 02127-1001 (50)

RIVETS, Stainless Steel

American Rivet Co., Inc., 11300 W. Melrose St., Franklin Park, IL 60131-1323 (6-1/2)
Banner Screw Co., Inc., 2901 W. Montrose Ave., Chicago, IL 60618-1403 (2)
Brainard Rivet Townsend/Textron, 222 Harry St., Box 30, Girard, OH 44420-1759 (6-1/2)
Celus Fasteners Manufacturing Inc., 2 Connector Rd., Andover, MA 01810-5904, (Also steel)
Conklin Brass & Copper Co., Inc., the T. E., 345 Hudson St., New York, NY 10014-3963
Dayton Electric Manufacturing Co., 5959 W. Howard St., Chicago, IL 60648-4014 (10)
Federal Screw & Supply Corp., 525 Broome St., New York, NY 10013-1647 (6)
Goodrich Co., B.F., Engineered Products Group, 500 S. Main St., Akron, OH 44318 (100)
Hi-Shear Corp., 2600 Skypark Dr., Torrance, CA 90505-5314 (99)
Mid-Continent Screw Products Co., 3701 W. Lunt Ave., Chicago, IL 60645-2615 (10)
Milford Rivet & Machine Co., 857 Bridgeport Ave., Milford, CT 06460-3139 (10)
Schnitzer Alloy Products Co., New Point Industrial Center, Box 433, Elizabeth, NJ 07206 (100)
Standard Optical Manufacturing Co., 42 Okner Pkwy., Livingston, NJ 07039-1604, (For eyeglasses) (10)
Standard Rivet Co., 71 A St., S. Boston, MA 02127-1001 (50)
USM Rivet Div., Emhart Corp., 510 River Rd., Shelton, CT 06484-4517 (100)

RIVETS, Tapped

American Rivet Co., Inc., 11300 W. Melrose St., Franklin Park, IL 60131-1323 (6-1/2)
Fischer Special Manufacturing Co., 111 Industrial Rd., Box 76500, Newport, KY 41076-9749 (10)
Milford Rivet & Machine Co., 857 Bridgeport Ave., Milford, CT 06460-3139 (10)

RIVETS, Threaded

American Rivet Co., Inc., 11300 W. Melrose St., Franklin Park, IL 60131-1323 (6-1/2)
CWR Manufacturing Co., 6424 Taft Rd., Box 2669, Syracuse, NY 13220 (30)
Urania Engineering Co., Inc., 198 S. Poplar St., Hazleton, PA 18201-7161

RIVETS, Tinners'

Albany Products Co., Inc., 150 S. New Boston Pk., Woburn, MA 01801-6204, (Aluminum, stainless, brass, bronze, monel, nylon) (25)

Assembly-Line Products Inc., Fastener Manufacturing Div., 44 Plaza Dr., Westmont, IL 60559-1130
Banner Screw Co., Inc., 2901 W. Montrose Ave., Chicago, IL 60618-1403 (2)
Brainard Rivet Townsend/Textron, 222 Harry St., Box 30, Girard, OH 44420-1759 (6-1/2)
Cobb & Drew Inc., Box 3387, Plymouth Center, MA 02361, (Brass, copper & steel) (3/4)
Reed & Prince Manufacturing Co., 1 Duncan Ave., Worcester, MA 01603-2301 (10)
Schnitzer Alloy Products Co., New Point Industrial Center, Box 433, Elizabeth, NJ 07206, (Stainless steel) (100)

RIVETS, Tubular

CHASE & CHUDIK INC., 1537 N. 25th Ave., Melrose Park, IL 60160, 312-345-0335; FAX: 312-345-9462 (2)
Reader Service No. 1705

STIMPSON CO., INC., 900 Sylvan Ave., Bayport, NY 11705-1097, 516-472-2000. For local phone numbers, see our listing in Corporate Index (5)
See our catalog pages in Corporate Index
Reader Service No. 760

American Rivet Co., Inc., 11300 W. Melrose St., Franklin Park, IL 60131-1323, (Drilled, etc.) (6-1/2)
Arrow Rivet Co., Inc., 60 Columbia Terrace, Braintree, MA 02184-1347, (All metals) (1)
Banner Screw Co., Inc., 2901 W. Montrose Ave., Chicago, IL 60618-1403 (2)
Century Fasteners & Machine Co., Inc., 4155 N. Rockwell St., Chicago, IL 60618-2822, (Tubular & solid)
Detroit Tubular Rivet Inc., 1213 Grove, Wyandotte, MI 48192-7045, (Solid, shoulder, brake & clutch)
Jacobson Group, Box J-2, Kenilworth, NJ 07033
Keystone Screw Corp., 535 Davisville Rd., Box V, Willow Grove, PA 19090-1525, (Semi-tubular, automotive) (10)
Lewis Screw Co., 4300 S. Racine, Chicago, IL 60609-3320 (7)
Metal Fastener & Bolt Co., 131 Belmont Ave., Toledo, OH 43602 (10)
Milford Rivet & Machine Co., 857 Bridgeport Ave., Milford, CT 06460-3139 (10)
Philadelphia Rivet Co., 266 N. Broad St., Doylestown, PA 18901-3426
Prairie Tubular Rivet Corp., No. Margaret St., Markesan, WI 53946
TCR Corp., 1600A 67th Ave., N., Minneapolis, MN 55430-1702 (54)

RIVETS, Wire

Brainard Rivet Townsend/Textron, 222 Harry St., Box 30, Girard, OH 44420-1759 (6-1/2)

RIVETS, Zinc

Brainard Rivet Townsend/Textron, 222 Harry St., Box 30, Girard, OH 44420-1759 (6-1/2)
Hassall Inc., John, Cantiague Rd., Westbury, NY 11590 (10)

ROASTERS (Nuts, Seeds, Grains, etc.)

National Drying Machinery Co., The, 2755 N. Hancock St., Philadelphia, PA 19133-3504 (9)
Stein Assocs. Inc., Sam, 1622 First St., C.N. 5001, Sandusky, OH 44870-3902 (109)

ROADBUILDING Machinery

American Lava Coatings Corp., 4 Oval Dr., Central Islip, NY 11722-1403, (Patching materials)
Blaw-Knox Construction Equipment Co., Rte. 16 E., Mattoon, IL 61938, (Pavers, wideners, spreaders, etc.) (100)
Chausse Manufacturing Co., 8100 Joy Rd., Detroit, MI 48204-3207, (Asphalt road maintenance equipment) (7)
Essick/Hadco Manufacturing, Div. of Figgie International Inc., 1950 Santa Fe Ave., Los Angeles, CA 90021-2925 (100)
Etnyre & Co., E.D., 200 Jefferson St., Oregon, IL 61061-1611 (1)
Flow-Boy Manufacturing, Box 1369, Norman, OK 73070-1369, (Horizontal discharge trailer)
Jeter Construction Co., Inc., Hwy. 21, Beaufort, SC 29902, (Paving)
Radiation Technology Inc., 108 Lake Denmark Rd., Box 185, Rockaway, NJ 07866-0001, (Stripping compounds) (7)
Redland Prismo Corp., 300 Lanidex Plaza, Parsippany, NJ 07054-2723, (Striping & marking equipment & materials) (80)
Southwest Welding Manufacturing, Box 5100, Long Beach, CA 90805-0100, (Construction)
Taylor Contractors Inc., N.E., 209 N. Rte. 50, Easton, MD 21601, (Excavating, ponds, drainage)
Wylie Manufacturing Co., A Sub. of E.D. Etnyre & Co., 200 Jefferson St., Box 72, Oregon, IL 61061-1611

Roasters, Rotary: *See Calciners, Rotary*

ROBOTIC Components

PARKER HANNIFIN CORP., PACKING DIV., 2220 S. 3600 West, Box 30505, Salt Lake City, UT 84130-0505, 801-972-3000 (100)

STOCK DRIVE PRODUCTS, A DSG COMPANY, DIV. OF DESIGNATRONICS INC., 2101-M Jericho Turnpike, New Hyde Park, NY 11040, 516-328-0200, ext. 30 (100)
Small inch & metric automation and drive components, incl. bearings, belts, gears, chains, couplings, fasteners, gearheads, pulleys, shafts, springs, speed reducers, rotary actuators, grippers, end-effectors, vibration mounts, inclinometers. Custom design capabilities
Reader Service No. 1286

Stock Drive Products
GRIPPERS
NEW-FREE CATALOG
(516) 328-3300 or 328-0200 Ext:30
2101-M Jericho Tpke., New Hyde Park, NY 11040
TWX: 510-223-0642 FAX: 516-326-8827

Reader Service No. 1286

Adaptive Intelligence Corporation, 2944 Scott Boulevard, Santa Clara, CA 95054-3312
Automatix Inc., 1000 Tech Park Dr., Billerica, MA 01821-4127 (30)
Barnes Drill Co., 814-30 Chestnut St., Rockford, IL 61101, (And systems) (20)
Dunham Tool Co., Inc., The, Saw Mill Rd., New Fairfield, CT 06810 (3/4)
Expert-KUKA Inc., 40675 Mound Rd., Sterling Heights, MI 48078 (36-1/2)
Fraser Fabricating & Inc., 1696 Star Batt Dr., Rochester, MI 48063 (1)
Gem City Engineering Co., The, 1546 Stanley Ave., Box 1295, Dayton, OH 45401
Hudson Robotics, Inc., 44 Commerce St., Springfield, NJ 07081
Penn-Field Industries, 420 Station Rd., Box 31, Quakertown, PA 18951-0031
Precision Controls Corp., 8537 York Rd., Box 240235, Charlotte, NC 28210
Robotic Vision Systems, Inc., 425 Rabro Dr. E., Smithtown, NY 11788-4227
Shaum Manufacturing Co., 1127 N. Nappanee St., Elkhart, IN 46514-1735 (23)
UAS Automation Systems Inc., 761 Middle St., Bristol, CT 06010, (Special designed systems)
V.S.I. Automation Assembly Inc., 165 Park St., Troy, MI 48083-2770 (20)

ROBOTS

HOBART BROTHERS CO., Hobart Square, Troy, OH 45373, 513-332-4000 (50)
Reader Service No. 702

ROBOTIC ARC WELDING SYSTEMS

Select power sources, interfacing and welding wire for all major industrial welding robots — or gain single source dependability with a complete Hobart Motoman package including training, warranty, service, parts. System includes choice of electric robots, power sources, wire feeders, controllers, interface, filler metals, teaching pendant and automated engineering services.

Phone: **(513) 332-4000**

HOBART.
Hobart Brothers Co. Hobart Square. Troy. Ohio 45373

Reader Service No. 702

AK Stamping Co., Inc., 1159 Rte. 22, Westfield, NJ 07092-2808, (Industrial)
AMCAM/Commercial Cam, Emerson Electric Co., Spring Lane, Box 309, Farmington, CT 06032, (Pick & place) (10)
Adaptive Intelligence Corporation, 2944 Scott Boulevard, Santa Clara, CA 95054-3312
Ameco Corp., Box 385, Menomonee Falls, WI 53051-0385, (Automatic material handling systems)
Automated Assemblies Corp., 25 School St., Clinton, MA 01510-3419, (And systems)
Automatic Tool Co., Inc., 1233 Broadway Ct., Box 3146, Rockford, IL 61104-1415, (Pick & place) (2)
Automatix Inc., 1000 Tech Park Dr., Billerica, MA 01821-4127, (Computerized systems) (30)
Cayuga Manufacturing Corp., 4979 Lake Ave., Buffalo, NY 14219-1313, (Robotic systems) (15)
Cincinnati Milacron, 4701 Marburg Ave., Cincinnati, OH 45209-1025 (100)
Clinton Machine Inc., 1300 S. Main St., Ovid, MI 48866-9724, (Robotic systems)
Commercial Cam, Div. of Emerson Electric, 1444 S. Wolf Rd., Wheeling, IL 60090-6514, (Fixed sequence)
Conair Inc., Conair Bldg., Franklin, PA 16323, (For removing molded plastic parts) (50)
Cyclomatic Industries Inc., 8123 Miralzhi Dr., San Diego, CA 92126-4342, (Welding) (5)
Dynamac Inc., 410 Forest St., Marlborough, MA 01752-3002, (Robot applications)
Dynamco Inc., Air Controls Div., 2639 Manana Dr., Dallas, TX 75220-1301, (2 axes, 4 degrees of freedom, pneumatic)
Ex-Cell-O Corp., Packaging Systems Div., 850 Ladd Rd., Walled Lake, MI 48088-2502 (200)
Expert-KUKA Inc., 40675 Mound Rd., Sterling Heights, MI 48078 (36-1/2)
Fabcon Systems Inc., 1100 E. Washington, Freeland, MI 48623-9050 (1/2)
Fluid Kinetics Inc., 4868 Factory Dr., Fairfield, OH 45014, (Up to 10 axis, gantry types)
Gros-Ite Industries Inc., New Britain Ave., Farmington, CT 06032, (Robotic applications) (20)
Henderson Industries, H. F., 45 Fairfield Pl., Caldwell, NJ 07006-6206, (Custom robotic software & hardware)
Hi Tech Robotics Ltd., Wilton Metal Manufacuring Div., 1210 E. Ferry St., Box 923, Buffalo, NY 14211 (30)
Hitachi America Ltd., 950 Elm Ave., 100, San Bruno, CA 94066-3036, (Industrial, electric) (50)
J.R.S. Machine & Tool Sales Corp. of America, Box 52, 333 Hamilton Blvd., South Plainfield, NJ 07080 (5)
Keller Technology Corp., 2320 Military Rd., Tonawanda, NY 14150-6005, (Systems) (1)
Kevco Tool & Manufacturing Co., Inc., Box 398, Kendallville, IN 46755-0398, (Robot components, tooling) (15)
Midway Machine Co., 2324 University Ave., St. Paul, MN 55114-1802, (Industrial) (3)
Miller, Richard K., 498 S. Main St., Madison, GA 30650-1639, (Robotics)
Mobot Corp., 980 Buenos Ave., San Diego, CA 92110-3925, (Automatic parts handling modular) (100)
Morris Co., The Robert E., 17 Talcott Notch Rd., Farmington, CT 06032-1818, (Industrial) (34)
Nypro Inc., 101 Union St., Clinton, MA 01510-2908 (160)
Positech Corp., Rush Lake Rd., Laurens, IA 50554
Prab Conveyors Inc., 5944 E. Kilgore Rd., Kalamazoo, MI 49003, (Industrial) (15)
Prab Robots Inc., 6007 Sprinkle Rd., Box 2121, Kalamazoo, MI 49003
Precision Automation Co., 1841 Old Cuthbert Rd., Cherry Hill, NJ 08034-1415, (Custom) (20)
Progressive Blasting Systems, 4201 Patterson Ave., SE, Grand Rapids, MI 49508-4033, (CNC controlled blast guns, 4-9 axes) (10-1/2)

MacRae's Blue Book, 1989. Reprinted with permission of Business Research Publications, Inc.

By contrast, another well-known directory, *Sweet's Catalog File*, consists entirely of bound volumes of catalogs. It is issued in several series, depending upon the needs of its users as to the types of catalogs included.

Figure 25-2. *MacRae's Blue Book*—Individual Company Listing

INGERSOLL

Ingersoll is comprised of a group of independent companies that offer a broad spectrum of manufacturing expertise. Working in close partnership with customers, Ingersoll is dedicated to creating long-term custom solutions that improve performance, uptime, and overall productivity.

Ingersoll designs and builds custom tooling, machines, and systems that are applied to the production of metal parts (or their substitutes) in a wide range of workpiece sizes and production volumes. The company also brings flexibility to high-volume applications in increasingly higher levels.

THE INGERSOLL MILLING MACHINE COMPANY, 707 Fulton Ave., Rockford, IL 61103, Telephone: Use numbers listed. Telex 257427. Detroit office: Timberland Office Park, 5455 Corporate Dr., Suite 200, Troy, MI 48098. Telephone: 313/641-5995.

Special Machines Group. 815/987-6000. Computer-integrated *flexible manufacturing systems* specifically engineered to customer requirements. Capabilities of *all* Ingersoll companies utilized to create the optimum production-automation-information balance for each application.

Portal-type and horizontal traveling column and table-type milling/boring/drilling machines; 50 to 250 hp 3, 4 and 5-axis machining centers, die sinking systems, special-purpose machine tools.

Automated high and mid-volume manufacturing systems including in-line and rotary transfer machines, head exchanger systems, shuttle machines and stand-alone machines.

Parts, repairs, preventive maintenance, rebuilding, upgrading and retrofitting of Ingersoll machine tools.

Simultaneous Engineering. 815/987-6000. Specific strategies to optimize both product and process design and maximize productivity. Transitional Programs and Continuous Improvement provide on-going support and upgrading.

Ingersoll Composites. Composites manufacturing machinery including contouring tape layers and pultrusion and filament winding machines.

Contract Machining and Fabrication. 815/987-6934. Complete machining and weldment capabilities for work to 75 tons or more.

INGERSOLL CUTTING TOOL COMPANY, 505 Fulton Ave., Rockford, IL 61103. 815/987-6600. Telex 257427.

Standard cutting tools. MAX-I on-edge insert indexable carbide face mills, end mills, slotting cutters and gear cutters. MAX-I-PRO high-production face mills for cast iron. Indexable drills. Rotary toolholders.

Special cutting tools. Milling cutters; high-production tooling packages; crankshaft and camshaft milling and turning tools; indexable form tools; feedout heads; CNC multiple-operation tooling; line boring bars; modified drilling tools; cylinder bore cutters.

INGERSOLL GmbH (Inc.) 1303 Eddy Ave., Rockford, IL 61103. 815/654-5830. Telex 257427.

EDM machines— CNC planetary sink-type for small to largest dies, molds and production parts.

Abraders for electrode production and finishing.

Thread whirling systems for threading holes from 1 1/4-in. to 14-in. I.D.

Tool grinders for carbide milling cutters.

Replacement parts for Ingersoll inserted blade milling cutters.

WALDRICH SIEGEN, Ingersoll GmbH (Inc.), 1303 Eddy Ave., Rockford, IL 61103. 815/654-5830. Telex 257427.

Turning machines—heavy-duty engine and roll lathes (including CNC) with swings up to 200 in. and turning lengths to 100 ft.

Roll grinders—precision cylindrical grinding up to 8-ft. dia and 80-ft. length.

Roll texturing machines (electrical discharge and laser) for steel mill work rolls.

Special heavy-duty machine tools for milling and boring.

WALDRICH COBURG, Ingersoll GmbH (Inc.), 1303 Eddy Ave., Rockford, IL 61103. 815/654-5830. Telex 257427.

Extrusion screw milling machines—special-purpose NC machines for milling the screws of plastic injection molding machines. 20 to 40 hp. Screw diameter from 1 to 26 in.

Guideways and surface grinding machines—conventional and CNC machines for jigs, fixtures and machine tool guideways 8 to 75 hp. Grinding lengths to 600 in.

Fluted roll grinding machines—for rolls to 18-in. dia, 160-in. length.

Special slotting machines and precision thread peeling machines.

Ingersoll Bohle, Ingersoll GmbH (Inc.), 1303 Eddy Ave., Rockford, IL 61103. 815/654-5830. Telex 257427.

Machining centers—horizontal and vertical spindle, 20 to 75 hp. See ad under Machining Centers.

Flexible Manufacturing Cells incorporating Ingersoll Bohle machining centers.

Bed mills—20 and 30 hp; table sizes 600mm wide × 1.6 or 2m long.

Special helical milling machines for screw compressors, rotors, etc.

MacRae's Blue Book, 1989. Reprinted with permission of Business Research Publications, Inc.

Many directories of manufacturers exist, some of a very specialized nature. A number of trade periodicals regularly publish directories of manufacturers of specialized equipment that would be of interest to their readers, such as those involved in chemical engineering, welding, or materials handling.

Closely allied to manufacturers' catalogs are publications known as data compilations, which are prepared by publishers who analyze the catalogs from a number of manufacturing firms and print booklets listing the features of competing products. This service is quite valuable because it saves the time of the engineer or scientist seeking a product with particular operating characteristics. Rather than poring over the individual catalogs from a dozen manufacturers, the designer could examine the tabular comparisons of these companies' products. Obviously, the compilations become outdated and must be updated regularly in order to prove useful; most of these compilations are sold on a subscription basis. Also, there is no guarantee that every company whose product should be considered is always included in the list of companies used in preparing the data compilations.

Certain fields are more likely than others to be represented by data compilations. Semiconductor products, such as transistors and microchips, are favorite topics, along with switches, capacitors, and similar products used in electric circuits.

TYPICAL EXAMPLES OF SOURCES OF MANUFACTURERS' LITERATURE

MacRae's Blue Book. New York: Business Research Publications, Inc.; 1893– . Annual.
> This venerable reference tool provides much useful information about manufacturers and their products. It is also strong in listing manufacturers' representatives, who can often serve well as local sources of information. In recent years this volume has also begun to include pages of catalogs of participating manufacturers.

Sweet's Catalog File. New York: McGraw-Hill Information Systems Company; 1914– . Annual.
> Consists of bound volumes of manufacturers' catalogs, with each volume devoted to different types of products, such as machine tools, architectural materials, or plant engineering products.

Thomas Register of American Manufacturers and Thomas Register Catalog File. New York: Thomas Publishing Company; 1910– . Annual.
> Although the majority of volumes in an annual set consist of a directory of American manufacturers, in recent years other volumes have been added in which manufacturers who wish to pay a fee can have pages from their own brochures advertising their products included. This publication offers a large, comprehensive listing of manufacturers, including

a detailed index of product headings to help one locate the correct category.

REFERENCE

1. Norton, Bruce. Vendor catalogs in science/technical libraries: why—and how. *Science & Technology Libraries.* 10(4):31- 41; 1990 Summer.
Describes the services and products of Information Handling Services, which provides customers with microform copies of trade catalogs, offering a dozen or so available options. Sold by subscription, the client can pick a particular service offering certain types of catalogs, all carefully indexed.

Chapter 26
Newspapers and Newsletters

HIGHLIGHTS OF NEWSPAPERS AND NEWSLETTERS

Description: Newspapers and newsletters devoted to science or technology are quite similar to their nontechnical counterparts, differing essentially only in their contents.

Significance: Newspapers of all sorts have several characteristics in common—they can be current, they can be rather detailed in their coverage of a topic, and they can cover many topics in a particular issue. Whereas periodicals are seldom issued more frequently than once a week, newspapers can be published daily. Newsletters, generally issued less frequently than newspapers, pride themselves on presenting information that may not be published elsewhere for weeks or even months. The contents of both formats tend to become out of date rather quickly, presenting readers with little of permanent value.

Quantity: There are dozens of sci-tech oriented newspapers and scores of newsletters in these disciplines.

Physical Characteristics: The typical sci-tech newspaper is apt to be published in the familiar tabloid size. Along with news items, they contain advertisements, tables, statistics, editorials and guest columns. Sci-tech newsletters tend to be published in the familiar 8.5" x 11" style, similar to those devoted to business, social issues, or any of the many other topics to which newsletters are devoted. They rarely vary from a format of solid text paragraphs.

Availability: Most sci-tech newspapers and newsletters are available only by subscription, and it would be an unusual public or academic library that would subscribe to many such specialized publications. Most subscribers are either individual sci-tech professionals or the special libraries serving them.

Retrieval: One of the problems with sci-tech newspapers is that most of them are not indexed by any abstracting/indexing service, making retrieval of older data very difficult. Some publishers offer help for requests about previous news items, but this is not always the case. In the past, newsletters suffered from the same retrieval difficulties, but in recent years online databases devoted to industrial newsletters have been developed.

Intended Audience: These publications are aimed at practicing sci-tech professionals, ranging from laboratory helpers to researchers to executives. Thus some of the publications emphasize very technical topics while others may feature issues relating to business and management.

Scope: Sci-tech newspapers and newsletters most commonly emphasize a rather narrow topic, perhaps no broader than electronics or biotechnology, or sometimes even more limited in scope. Others combine technical material with business/management topics.

NATURE OF NEWSPAPERS AND NEWSLETTERS

There are scores of newspapers and newsletters devoted to selected areas of science and technology. They make no attempt to appeal to the general public; rather there is an emphasis on the topics that would interest professionals working in special sections of industry and research. Although the information these formats contain tends to be rather ephemeral, the need for very current news offsets their lack of long-term value.

Most of these newspapers and newsletters are geared to the needs and interests of people in the industrial or commercial segments of society rather than those in the scholarly sector. It is common for a commercial firm devoted to applied research and development to subscribe to scores of copies of one newspaper aimed at an appropriate type of industry, such as plastics or electronics.

One of the problems with newspapers is the lack of means to retrieve older news items. Most indexing and abstracting services do not index them. Short of setting up an index in an organization's library, the only source of help would be the publishers' records or indexes, if they exist. Because much of the news has a short time span of interest, the need for retrospective searching is not an everyday problem.

Eventually much of what these publications cover is published in other formats, such as industry-oriented periodicals or, rarely, in daily newspapers of general interest. The fact that Mr. XYZ has just been promoted to be the head of the semiconductor division of the ABC Corporation is not apt to appear in many other publications, whereas the first release of a new government regulation regarding the export of advanced computers is most likely to be published later in many other publications. *Chemical Marketing Reporter* is a typical sci-tech newspaper; it features not only news items about the industrial chemicals field but also statistics on the prices and output of these materials.

Other sci-tech newspapers tend to be broader in scope and less oriented towards technology. An example of this type of newspaper is *The Scientist*, which contains news items about scientific developments, interviews with scientists, book reviews, and notices of meetings.

Newsletters tend to be much smaller than newspapers, perhaps just a few pages stapled together. Their main value lies in the ability of their publishers to be the first into print with really important new developments. If this sort of information weren't regularly published in these newsletters before it became generally known, the success of such publications would be in jeopardy. They are generally edited by someone with a sufficiently strong background in the field to be called an expert. These editors have their own sources of information for learning quickly about new topics. A relatively recent addition to the ranks of newsletters is entitled *Science Watch*; its goal is to track trends and performance in basic research. Thus a typical issue would contain articles on the growth of new fields, such as the sample article found in Figure 26-1 on the progress of Japanese scientists in developing organic superconductors.

Another type of newsletter is published by professional organizations; typical contents include news items and announcements of interest to organization members.

At least two online databases have been created to remedy the lack of coverage of newsletters by traditional abstracting and indexing services. One, PTS NEWSLETTER DATABASE, began in 1988 and provides daily updates of the full text of more than 100 newsletters in such fields as biotechnology, aerospace, energy, environment, and computers.

TYPICAL EXAMPLES OF NEWSPAPERS AND NEWSLETTERS

Chemical Marketing Reporter. New York: Schnell Publishing Company; 1871– . Weekly.
This newspaper is a source of information about the costs of various chemicals, statistics on production of chemicals, news of personnel in the field, new products, and other topics related to the field of chemical production.

PTS NEWSLETTER DATABASE. Cleveland: Predicasts; 1988– . Updated daily.
This database regularly indexes the contents of more than 100 of the leading newsletters in such fields as air pollution, defense industry, oil industry, and biotechnology. The full text of articles is given, and the scope is international.

Figure 26-1. *Science Watch*—Sample Article

Japan Targets Organic Superconductors, Takes The Long View

"IBM used to be involved in a big way, but it wasn't making progress, so it greatly reduced its effort. So did Bell Labs, Xerox and others," says Jack Williams of Argonne National Laboratory, Argonne, Illinois, talking about recent participation by U.S. industrial firms in research on the other kind of superconductors—organic superconductors.

Meanwhile, Japanese researchers have steadily increased their presence in this field over the past five years. "It has become a targeted area for Japan," observes Williams.

And it is to a Japanese team led by Gunzi Saito at the Institute of Solid State Physics, University of Tokyo, that the garland goes for having achieved the highest superconducting transition temperature (Tc) yet in an organic compound, the sulfur-based salt BEDT-TTF, at ambient pressure (10.4 K). The report of this breakthrough ranked as the tenth most cited paper in the physical sciences over a recent two-month period (March-April 1989).

First discovered in 1980, organic superconductors have recently received considerably less attention than have their headline-grabbing copper oxide counterparts. In fact, although progress in increasing the transition temperatures in organic superconductors has been steadily achieved since the beginning of the decade, many researchers left this field in 1987 to pursue work on the new higher-temperature, ceramic-based superconductors. Some, including Williams and his group, have now returned to the organics because of their great potential for higher Tc's and easy chemical modification.

In many ways, the organic superconductors have much in common with the ceramics. They offer many of the same advantages for electronic and magnetic applications, but there are difficulties. "A lot of progress has yet to be made to get the organic superconductors processable," Martin Bryce, University of Durham, U.K., tells *Science Watch*. "They are still brittle, frail crystals. The fabrication problems are essentially the same as for the ceramics." Still, Bryce remains optimistic. "Prospects are good for the organics," he says.

Clearly the Japanese think so, too. Their presence in the field is "very strong," says Williams. That view is corroborated by data from the Institute for Scientific Information's 1988 Research Front Database. An analysis of the papers in the 1988 specialty area on organic superconductors revealed that nearly a third were by Japanese researchers. Not only did Japan produce more papers than any other nation (a 30.8% share of the 271 papers), but it outperformed itself, far surpassing its ex-

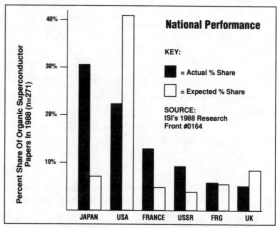

National Performance

KEY:

◼ = Actual % Share

☐ = Expected % Share

SOURCE:
ISI's 1988 Research
Front #0164

Percent Share Of Organic Superconductor Papers In 1988 (n=271)

JAPAN USA FRANCE USSR FRG UK

pected share (7.1%), which is the representation of Japanese papers in the database.

The United States was second in 1988, with a 22.5% share—about half its expected rate of participation.

Is the United States, then, behind Japan? Well, yes and no, say scientists with whom *Science Watch* spoke.

Fred Wudl of the Institute for Polymers and Organic Solids, University of California, Santa Barbara, hesitated to concede a scientific superiority, saying, "I'm not sure they are ahead, although the new record is held by the Japanese." He explained that much of the fundamental work on which the Japanese and others are now building was conducted

Institutional Standings

Rank	Institution	% 1988 Papers
1	University of Tokyo	10.3
2	USSR Academy of Sciences	5.0
3	Argonne National Lab	3.8
4	Electrotechnical Lab, Ibaraki	3.2
5	CENS (Saclay Nuclear Research Center), Gif-sur-Yvette	3.1
6	University of Durham	2.7
7	Osaka University	2.0
8	Institute for Molecular Sciences, Okazaki	1.9
9	University of Paris, XI	1.8
10	Hiroshima University	1.7

SOURCE: ISI's 1988 Research Front #0164.

by U.S., European, and Soviet scientists. In terms of sheer effort, however, Wudl, Bryce, and Williams were not surprised by the dominance of Japan in output.

In the United States, one of the main centers of organic superconductor research is Williams' group at Argonne. Wudl at Santa Barbara and Dwaine Cowan at Johns Hopkins University, Baltimore, are also key players.

But the biggest player is Saito at the University of Tokyo. Last year his team fielded one out of every ten papers on organic superconductors (see table). Recently in Tokyo, from August 23-26, Saito hosted the first international symposium devoted exclusively to organic superconductors.

Reinforcing the view of Japan's dominance in this area is the finding that five of the top ten institutions are Japanese. Of the U.S. scientists working on organic superconductors *Science Watch* spoke with, none complained of inadequate funding; a few thought they had in fact been well supported.

But the lack of participation by U.S. industrial firms is notable. Can this be a case study of how, more and more today, U.S. businesses seem unwilling to invest in anything that will not produce short-term returns? When U.S. companies headed for the exits at mid-decade, the Japanese, taking the long view, entered and invested their resources. Japan now seems willing to wait patiently for its returns—possibly in the next decade or even beyond. And, in the meantime, its scientists are, in the words of Wudl, "throwing everything but the kitchen sink at the problem." ◼

Science Watch. Philadelphia: Institute for Scientific Information; 1989– . Monthly.

Concentrates on new developments in basic research, often featuring tables and charts to illustrate its articles. Aimed at scientists in basic and applied fields.

The Scientist. Philadelphia: Institute for Scientific Information; 1986– . Biweekly.

This newspaper contains current news items of interest to the science professional, as well as interviews, book reviews, software reviews, classified ads, and information about upcoming meetings, fellowships, and research opportunities.

Chapter 27
Reviews of the Literature

HIGHLIGHTS OF REVIEWS OF THE LITERATURE

Description: Reviews of the literature consist of surveys of the important publications on a selected topic that were usually written during a specific time period and are published as books or articles.

Significance: Carefully selected publications which represent the highlights of what was published on a particular topic can be of great value to scientists and engineers, saving them the time and trouble of examining a large amount of literature in their search for the most significant items. Reviews of the literature provide their readers with such summaries. In some instances the reviews are not limited to literature that was published in a particular time period.

Quantity: Several dozen annual reviews of the literature are published as separate books, often in annual series that have been issued for several years. Other sources for reviews are lengthy periodical articles devoted to this type of summary; scores of review articles are written for scholarly journals each year.

Physical Characteristics: Reviews of the literature that are in book form tend to look like regular monographs except that they usually have a half dozen or more chapters, each on a separate but related topic, that have long bibliographies. On the other hand, some reviews appear in periodicals and look like typical long periodical articles.

Availability: Reviews are available in libraries collecting research materials in the subjects involved; few libraries would be apt to contain all the annual series that exist.

Retrieval: There are at least three series titles commonly used for the hardbound annual reviews; the titles are easily located in a library catalog: *Annual Reviews of . . .* , *Progress in . . .* , and *Advances in. . . .* Another way to locate them is by means of the sub-heading "—Reviews" that might appear under any particular topic. At least one indexing and abstracting service is devoted to review literature.

Intended Audience: Most review literature is written for scholarly practitioners and graduate students.

Scope: Annual reviews in book form usually contain enough material to justify saying that such titles cover broad subjects, like organic chemistry, biology or astronomy. Journal articles, on the other hand, are limited to much smaller topics.

NATURE OF REVIEWS OF THE LITERATURE

In view of the great quantity of sci-tech information being generated each day, it is becoming increasingly difficult for an individual to keep abreast of new developments. Even when a person's interests may be quite narrow, the literature is literally being created faster than one can read it. Because of this situation, it is fortunate that a particular kind of sci-tech information source exists which can help solve this dilemma, namely reviews of the literature.

Primarily a product of the past 40 years, reviews of the literature have been published on a variety of subjects. They vary in complexity, scope, and frequency of publication, but they all serve the purpose of providing a survey of the best material on a given topic, thus saving an interested scientist or engineer from having to go through a laborious process of selecting and evaluating pertinent items. These reviews are not confined merely to listing the key items; their main function is to analyze the significance of the literature, showing trends in the quantity and nature of what has been published, and perhaps commenting on what was not published.

Reviews of the literature have generally concentrated on analyzing periodical articles, with monographs considered of secondary importance. The compilers, almost always experienced practitioners or scholars in a special field, must examine large amounts of book and periodical literature in order to select the most significant items. A typical review of this sort would include hundreds of citations to the literature, each evaluated by the compiler. Compiling such a review is a tremendously time-consuming task for a compiler; reviews of the literature make it unnecessary for every interested reader to duplicate this investment of time and effort.

Journals created for the express purpose of reviewing the scientific literature have a long and useful history. David Kronick states that one such journal began its life in Leipzig around 1750,[1] but the growth of review journals was slow thereafter. Despite the long history of a few special journals aimed at reviewing the literature, for centuries most reviews of the literature appeared as articles in journals, particularly scholarly journals.

In the 1940s an important variation began to appear in regard to the type of publication in which these review articles were published.

Reviews, rather than being published as articles in particular periodicals, began to appear as bound annual serials developed solely for that purpose. At first, the reviews attempted to summarize developments occurring during a recent calendar year. As time went on, more and more such series began publication, and less importance was given to the publication date of items selected. Now most annual review series make no attempt to confine the literature being analyzed to a particular year of publication.

Although the number of such annual series is no longer increasing at the rapid rate maintained in the 1950s and 1960s, scores of them are still being published annually. There are three titles which have proved most popular among publishers of such bound serials, namely *Advances in . . . , Progress in . . . ,* and *Annual Review of. . . .* An interesting example of an annual series that turned into a review journal is *Progress in Surface Science.* Its first volume appeared in the traditional bound form used for annual publications, but in its second volume and subsequent volumes appeared as a journal, with three/four issues per year.

Typical subjects for review series include chemical engineering, nuclear science, and oceanography. A collection officer responsible for selecting materials on a particular subject generally feels obliged to buy the series closest in subject coverage to the main thrust of the collection, especially in libraries with research collections.

For many years, one of the problems with the annual review series, in addition to the cost they added to the serials budgets of libraries, was the fact that they were not consistently indexed by traditional abstracting and indexing services. In the eyes of these services the annual review series was difficult to categorize—they weren't monographs and they weren't periodicals. One of the best solutions to this problem was the creation of a special index for all sorts of review literature—periodical articles as well as annual review series. In 1974 the Institute for Scientific Information created its *Index to Scientific Reviews,* which provides a semiannual index (with an annual cumulation) to this literature. It has remained a key tool for searching reviews, although other indexing services have since begun to index certain review series.

TYPICAL EXAMPLES OF REVIEWS OF THE LITERATURE

Index to Scientific Reviews. Philadephia: Institute for Scientific Information; 1974– . Semiannual with annual cumulations.
> Indexes some 30,000 review articles from 3,400 periodicals and annual review publications, covering all disciplines of science and engineering. Allows for searching by authors, title words, and author affiliations. Includes a citation index, in which references cited at the end of a publication are searchable by authors' names.

Progress in Surface Science. New York: Pergamon Press; 1972– . Quarterly.

After one year as an annual bound volume, it became a journal with three to four issues per year. It deals with such topics as semiconductors, films, and spectroscopy. Typical issues contain one or two articles, often with over 100 citations to the literature for each paper.

REFERENCE

1. Kronick, David A. *A history of scientific and technical periodicals: the origins and development of the scientific and technical press, 1665-1790.* 2d ed. Metuchen, NJ: Scarecrow Press; 1976: p. 184-201.

An excellent analysis of the growth and significance of sci-tech publications, including indexes covering them.

Chapter 28
Standards and Specifications

HIGHLIGHTS OF STANDARDS AND SPECIFICATIONS

Description: Standards and specifications are documents that describe a prescribed way of performing a process or making a product. They represent agreements reached by groups authorized to produce these documents.

Significance: Standards and specifications serve several vital functions. They help ensure that products and processes will provide maximum safety and adequate quality to users. They help ensure proper performance of the product or process. Finally, they help ensure that one manufacturer's output will be compatible with those of other manufacturers.

Quantity: Thousands of specifications and standards are created by government agencies, private companies, and professional organizations.

Physical Characteristics: Standards and specifications are usually paperbound documents that may contain charts, tables, and illustrations along with a tersely written text.

Availability: Most standards and specifications are available from the issuing sources. There are also private companies that sell them, either on a subscription basis or in single copies.

Retrieval: The usual method for identifying a given standard or specification is to use the indexes provided by the issuing body. However, in recent years online indexes to several types of standards have appeared.

Intended Audience: Standards and specifications are intended for use by people involved in the design and preparation of certain products and are not for the general public.

Scope: Standards and specifications are usually narrow in scope. While standards tend to be broader than specifications, neither of them could be called broad-based. A standard might prescribe the way in which all types of airborne electronic devices must be constructed, while a specification might describe the proper way to construct a particular airborne product, such as a radio designed for use in a military aircraft.

NATURE OF STANDARDS AND SPECIFICATIONS

Standards and specifications can be defined as prescribed ways of performing certain processes or making certain products. It takes only a moment's reflection to realize that we are all living in the midst of a society that is greatly influenced by standards. For example, the electric traffic lights so familiar to us universally use a red light for stop and a green light for go. Any effort to reverse this practice would be chaotic. In the United States we drive on the right-hand side of roads, while in Great Britain the left side is used. As long as drivers know which is correct, there are no problems. Getting into more intricate examples, a laboratory thermometer that reads a certain degree would give the same reading for that same amount of heat in any country of the world because the same scale for thermometers is universally accepted. Again, it would be chaotic if scientists (and manufacturers of thermometers) had not agreed on a standard.

Another benefit provided by standards and specifications is the interchangeability of one manufacturer's output with that of another. It would be an impossible situation if the makers of lamp sockets used a design that only one brand of light bulb would fit. Consumers also benefit from lowered costs when the number of variant products is reduced by agreed upon standards. Manufacturers of light bulbs, for example, don't have to make ordinary bulbs for a particular wattage rating with one sort of threads on the base and other bulbs of the same wattage with a different type of base.

Safety is another by-product of the creation of standards and specifications. Without standards carefully prepared by skilled engineers, a product would be more likely to fail in some situations. A seat belt in a car might not be strong enough unless manufactured according to standards.

Despite the benefits that standards and specifications provide to the general public, they are not written to be read by the average person. They are intended for use by the engineers and designers preparing certain products; they usually contain very specific technical terms that would be meaningless to the average person.

Standards and specifications as a distinct type of literature play an important role in certain kinds of sci-tech libraries. *Role of Standards in Sci-Tech Libraries*, edited by Ellis Mount, contains accounts of the types of standards collections maintained in various types of sci-tech libraries, as well as services associated with such materials.[1]

Generally speaking, standards are most important in engineering and applied science installations, where new products are being developed. They are particularly important in companies manufacuturing products on government contracts. For example, an engineer designing a new and improved model of an aircraft engine must be very much aware of the requirements laid down by government and/or commercial organizations for such engines. His or her company would be wasting its money if it developed a new engine that didn't meet required standards. On the other hand, a research chemist studying the effects of one chemical on another is not concerned with a standard way of producing some new product because the stage of the research is too far removed from the need to standardize a product—he or she may not even have a worthwhile product to standardize for all that is known during the early months of experimentation.

There are no official definitions of the differences between standards and specifications, but general practices do give a clue as to their characteristics. Quite often, standards have rather broad coverage compared to specifications. For example, a state highway department would have several standards for the bridges it builds; the standards might cover such major points as the relation of the expected load on the bridge to the maximum load the bridge could safely carry, or on the width of traffic lanes crossing the bridge. On the other hand, specifications might be different for each bridge, such as the number of lanes of traffic in each direction, the type of pedestrian walkway (if any), or the kind of lighting on the bridge (if any).

Probably the largest single source of standards in this country is the federal government. It publishes two series of specifications and standards: those developed by the Department of Defense and those developed by nonmilitary agencies, such as the General Services Administration. Examples of indexes issued by these departments are listed at the end of this chapter. They are cumulated annually, with monthly updates.

Having one agency in the Department of Defense supervising the preparation of military standards eliminates a lot of duplicated effort among the Army, Navy, Marines, and Air Force to develop similar standards. This cooperative system was established after World War II when it became obvious that much money was being wasted by having each service develop its own standards.

Military standards are denoted by the prefix "MIL" in the numbers of the documents, such as MIL Std 123, or MIL E-5444k (these are not actual numbers). The former designation would be given to a rather general publication, such as one defining the features required for all airborne gear, while MIL E-5444 might be more specialized and refer only to the features or requirements for a particular radar set. Letters following the numbers indicate the edition of the document; it is vital in working with standards and specifications to be sure that the latest edition is being used. A page from a typical military specification may be found in Figure 28-1.

Nonmilitary types of standards and specifications from the federal government are called federal standards, to distinguish them from the MIL standard system. Federal standards and specifications cover a multitude of subjects, ranging from canned food for government cafeterias and restaurants to metal office desks. A typical number (imaginary) might be LL-KKK-27g, where the final letter again represents the edition number.

As the listings at the end of this chapter indicate, both kinds of government standards are listed in annual summaries, with updates issued periodically.

Outside of the federal government, one of the largest sources of standards is the American National Standards Institute (ANSI), a private organization that is authorized by federal law to coordinate the issuance of standards by professional societies in the United States as well as to sponsor committees to create standards not covered by professional societies. ANSI standards range from those for preparing bibliographic citations to proper ways to prepare blacktop mixes for streets and highways. Many engineering organizations and groups of manufacturers prepare pertinent standards for their work, although most are given ANSI numbers and are sold by ANSI to the general public. ANSI issues an annual catalog of available standards.

Several ANSI standards are related to libraries and information services, such as the *American National Standard for Bibliographic References*, ANSI Z39.29-1977, currently being updated. It provides guidelines for the creation of citations to books, journal articles, slides, computerized records, and other informational formats. There are several other ANSI standards bearing on computers and publications.

Still other sources of standards are professional organizations, such as the American Society for Testing and Materials. ASTM specifications cover all sorts of materials, from concrete mixes for highways to constituents needed in certain types of steel.

Household products such as stoves and electrical devices are checked by a private organization known as the Underwriters Laboratory. The UL mark on a product is proof that it has been constructed according to Underwriters Laboratory standards, a guarantee of safe operation and dependable service.

Practically every manufacturing plant of any size will have its own standards department in which standards required for producing a particular product or process are developed as needed. There is a profession of standards engineers whose careers are devoted to the creation of standards.

Figure 28-1. Federal Military Specification

```
                                              MIL-E-5400J
                                              7 DECEMBER 1986

                                              Superseding
                                              MIL-E-5400H
                                              1 June 1965
                                              Supersession data
                                              (see 6.8)

                      MILITARY SPECIFICATION

        ELECTRONIC EQUIPMENT, AIRCRAFT, GENERAL SPECIFICATION FOR

        This specification is mandatory for use by all Departments and Agencies
                    of the Department of Defense.

1.  SCOPE

    1.1  Scope.- This specification covers the general requirements for the
design and manufacture of airborne electronic equipment for operation primarily
in piloted aircraft.  The detail performance and test requirements for a particular
equipment shall be as specified in the detail specification for that equipment.

    1.2  Classification.- The electronic equipment for which the general require-
ments for design and manufacture are outlined shall be of the following classes,
as specified (see 6.2):

        Class 1   - Equipment designed for 50,000 ft. altitude and continuous
                    sea level operation over the temperature range of -54°
                    to +55° C.
        Class 1A  - Equipment designed for 30,000 ft. altitude and continuous
                    sea level operation over the temperature range of -54°
                    to +55° C.
        Class 2   - Equipment designed for 70,000 ft. altitude and continuous
                    sea level operation over the temperature range of -54°
                    to +71° C.
        Class 3   - Equipment designed for 100,000 ft. altitude and continuous
                    sea level operation over the temperature range of -54°
                    to +95° C.
        Class 4   - Equipment designed for 100,000 ft. altitude and continuous
                    sea level operation over the temperature range of -54°
                    to +125° C.

    The addition of the letter "X" after the class number, e.g., (class 1X), will
identify the equipment using forced air or other types of auxiliary cooling.

2.  APPLICABLE DOCUMENTS

    2.1  The documents of the issue listed in the following bulletin, of the
issue of the bulletin 1/ in effect on date of invitation for bids  or request for
proposal, form a part of this specification:

1/ Later revisions, amendments, Qualified Products Lists, and superseding docu-
ments may apply when preferred by the contractor.

                                              FSC-MISC
```

On the international scene, the International Standards Organization has developed a large number of standards, and American efforts are coordinated so that U.S. standards are at least in general agreement with ISO standards, if not very closely tied to them.

TYPICAL EXAMPLES OF INDEXES OF STANDARDS AND SPECIFICATIONS

American National Standards Institute. *Catalog.* New York: The Institute; 1948– . Annual.
Lists all the specifications and standards which were approved by ANSI, including those from special ANSI committees as well as from professional organizations. Includes a subject index.

U.S. Department of Defense. *Index of Specifications and Standards.* Washington, DC: Government Printing Office; 1951– . Bimonthly.
This is the official listing of military specifications and standards. There is an annual cumulation; a subject index is provided.

U.S. General Services Administration. *Index of Federal Specifications and Standards.* Washington, DC: Government Printing Office; 1952– . Monthly.
Lists the nonmilitary specifications and standards of the federal government. Includes a subject index and cumulates annually.

REFERENCE

1. Mount, Ellis, ed. *Role of standards in sci-tech libraries.* New York: Haworth Press; 1990. 127 p. (Also published as vol. 10 no. 3 of *Science & Technology Libraries.*)
Presents half a dozen articles about the use, types, and sources of standards in various sorts of sci-tech libraries, including corporate, public, government, and academic units. Also discusses the role of commercial sources of standards.

Chapter 29
Tables

HIGHLIGHTS OF TABLES

Description: Tables consist of numerical data, arranged for quick access by readers. They may appear as part of a larger work, such as a book or periodical article, or as part of a collection of other tables. This chapter is chiefly concerned with the latter case.

Significance: Many times a scientist or engineer needs to find a particular number, such as the boiling point of a chemical, and needs to find such data quickly. Tables, often found in bound collections containing many tables, can fit that need, usually allowing speedy access. However, tabular data, like any other kind, can get out of date, in which case sources other than bound collections must be used, such as the many periodical articles providing such data.

Quantity: There are hundreds of bound collections of sci-tech data in the form of tables, some general and some very specific. There are thousands of periodical articles that serve to provide current, specific tabular data.

Physical Characteristics: Bound collections of tables may be arranged by broad subjects, with a more detailed subject approach provided by the index of the book. Many of them have the appearance of handbooks. Periodical articles devoted to tabular data have no distinguishing marks to identify them.

Availability: Most libraries place bound collections of tables in their reference section, unless a particular volume is so infrequently used it can be shelved with circulating books.

Retrieval: Some sources may have an entry under the term "Tables" or perhaps under the name of a particular discipline followed by the sub-heading "Tables" (such as Thermodynamics—Tables). Library catalogs and periodical indexing services both provide access to such materials.

Intended Audience: Audiences for tables can vary from high school students to practicing scientists and engineers. However, the more theoretical a project or study the less likely tables would be needed. A scientist engaged in formulating a basic concept of some sort would be unlikely to need specific data such as that found in the usual table. Engineers are more apt to need tables than scientists.

Scope: Collections of tables can vary greatly from those covering all aspects of a discipline, such as chemistry, to those limited to a specific discipline, such as electroplating or colloids. The more specific the topic, the more likely it is that the main users would be experienced people, rather than students. The tables found in periodical articles are more apt to appeal to the more advanced person than to students.

NATURE OF TABLES

Much of the information required by scientists and engineers is found in tabular form rather than in descriptive formats (such as articles and texts). Tables contain a variety of data, such as measurements of states of matter (boiling points, yield strengths), mathematical values (logarithms, trigonometric functions), or even what could be called routine statistics (number of chemists employed in a given year, size of budgets for research in companies). As can be seen from these brief examples, tabular data vary greatly in their nature, complexity, and field in which needed. It would be hard to conceive of tabular data being at all important in the humanities or the arts, whereas there is a high degree of interest in such information in most sci-tech disciplines.

Tabular data probably appear most commonly in book form; the books often consist of collections of tables (such as a collection of mathematical tables) or may be devoted to a single topic (such as the properties of alloys). Some tables are found in periodical articles, usually centered on a specific topic. Other sources of tables are technical reports and government documents.

Tables must be prepared with a high level of accuracy in order to be considered as valid sources for scientists and engineers to use, and most tables do meet this requirement. One of the big problems with tabular data is keeping them up to date. Although mathematical tables do not get out of date (the logarithm of a number isn't going to change over the years), tabular data concerning physical objects do change in time. For example, new methods of determining boiling points of liquids become more accurate as new equipment and techniques are developed. Therefore, publishers are faced with the need for periodic updating of books; periodical articles, on the other hand, are simpler to revise as needed.

One benefit of the computer age in which we live is the simplification of calculation of values depending upon equations. Proper programming can produce huge amounts of data representing different values inserted in sci-tech equations. Thus errors of calculation and the cost of compiling tables should both diminish through computerization. Benefits might include the cheaper price of books of tables as well as the appearance of books in tabular form that might not have been economically feasible without computers to calculate and print them.

Still another value of computers is the inclusion of tabular data in machine-readable form, including CD-ROMs and reels of magnetic tape. Such formats allow interested sci-tech users to study the data on their own computers, providing speed and flexibility of use.

One of the problems of using tables is finding available tables quickly. Over the years books have been published that were designed as indexes to what tables were available. A notable example is *An Index of Mathematical Tables* by A. Fletcher et al., which set the standard for the ideal guide to mathematical tables. Although now out of date, the *Index* has in a sense been updated by certain journals which are generally confined to listing articles describing new tables. More recently, online searching has made it simple to locate literature which include tabular data, a step toward resolving the problem.

One of the best-known compilations of tabular data is the *Handbook of Chemistry and Physics*. This annual volume contains hundreds of tables devoted to chemistry, physics, mathematics, and a number of other fields. It is used by scientists, engineers, and students; its annual updating helps keep it abreast of the times.

Closely related to tables are data compilations, which consist of facts and figures about particular commercial products, arranged more or less in tabular form. More information about such publications may be found in Chapter 25, "Manufacturers' Literature."

TYPICAL EXAMPLE OF AN INDEX TO TABLES

Fletcher, A. and others. *An index of mathematical tables.* 2d ed.
Reading, MA: Addison Wesley; 1962. 2 vols.
Although now out of print, this is a very useful guide to mathematical tables, whether in book form or in periodical articles. It contains a detailed index and a description of each table. Over the years the descriptions of tables found in this book have been supplemented by those found in a journal entitled *Mathematics of Computation* (formerly known as *Mathematical Tables and Other Aids to Computation);* this journal lists both published and unpublished tables.

TYPICAL EXAMPLE OF A COLLECTION OF TABLES

Handbook of Chemistry and Physics. Boca Raton, FL: CRC Press; 1913– . Annual.

This is probably the best-known collection of sci-tech tables. Devoted to chemistry and physics, but with a strong section for mathematics, it consists entirely of tabular data covering all aspects of these disciplines. Tables are arranged by broad categories, and in addition there is a detailed subject index. Each new edition includes new tables and revisions of many others.

Chapter 30
Taxonomic Literature

HIGHLIGHTS OF TAXONOMIC LITERATURE

Description: The literature that deals with handling, identification, and classification of specimens is called taxonomic literature.

Significance: Published descriptions of species enable researchers to identify specimens they have without seeing the original examples of the species. Taxonomic literature provides structured descriptions of all known species. It is very important in evolutionary and ecological research, for example.

Quantity: Articles, pamphlets, and monographs abound and can be found for all types of flora and fauna.

Physical Characteristics: Journal articles and pamphlets often describe a particular species of plant or animal, whereas monographs usually cover entire families of organisms.

Availability: Well-known classes of organisms are covered in a variety of readily-available resources, but it is much more difficult to identify literature for lesser-known organisms.

Retrieval: Indexes and abstracting services can be used to identify journal articles; standard publishing sources will identify monographs. Pamphlets are difficult to find; some of them may be located through library catalogs.

Intended Audience: Taxonomic literature can be used by anyone who wishes to identify a specimen, whether they are novices or experts.

Scope: All organisms are classified. These listings and descriptions cover phylum (or division), class, order, family, genus, and species.

NATURE OF TAXONOMIC LITERATURE

All organisms and specimens in zoology, biology, and botany are classified into categories known as kingdoms, phyla (or, in the case of botany, divisions), classes, orders, families, genera, and species. The classification schemes allow researchers to identify specimens they have discovered and to know what type of organism they have found. These schemes also enable researchers to identify new species which have not previously been described.

Of utmost importance in the discovery and naming of hitherto unknown, or unidentified, specimens is a structured method of devising nomenclature and descriptions understandable to others. A properly named and described specimen can be recognized by others who find it. Precision and consistency are very important. To provide this precision and consistency, taxa (singular form is taxon) were developed. A taxon is a group of organisms sharing several characteristics, or a formal unit at any level in a hierarchical system sufficiently distinct to be worthy of bearing a name.[1]

There are several interpretations of the field of taxonomy. Some authors call it the "science of classification,"[2] while others describe taxonomy as one component of the discipline known as "systematics,"[3] which includes classification and all other aspects of dealing with organisms. Regardless of the interpretation of the term, all agree that the objectives of taxonomy are:

1. To discriminate among organisms and to provide means for subsequent recognition (identification) of the discriminated entities (taxa)
2. To develop a suitable procedure (nomenclature) for designating taxa for reference purposes
3. To devise and perfect a scheme of classification in which the named taxa can be arranged.[4]

Therefore, any person involved in the discovery of new animal, plant, or microbial specimens must have taxonomic literature available to them. Such literature varies greatly. There are monographs that list general descriptions of all taxa, and there are pamphlets or periodical articles that identify, in detail, very specific genera and species of organisms. One can find descriptions of a particular entity, or revisions of previously identified classifications, or complete and extensive monographs on species. There are field guidebooks, catalogs, and lists of specimens.

All of these publications usually "consist of descriptions, diagnoses, keys, synonymies, and diagrammatic representations of the relationships of taxa. Keys present a series of choices, usually dichotomous, leading to the correct name of a species. Key characters are those attributes which readily distinguish between species or higher taxa. Synonymies are lists of names that have been applied to a particular taxon. . . . Diagrams are often used to represent the author's views of relationships. They may be in the form of a two-dimensional 'tree,' called a dendrogram, or they may employ perspective and thus represent the facts in three dimensions."[5]

TYPICAL EXAMPLES OF TAXONOMIC LITERATURE

International code of zoological nomenclature. 3d ed. Published for the International Commission on Zoological Nomenclature by the International Trust for Zoological Nomenclature in Association with the British Museum (Natural History). Berkeley: University of California Press; 1985. 338 p.
Adopted by the 10th General Assembly of the International Union of Biological Sciences, the *Code* is the work of a committee with members in Australia, France, the United Kingdom, and the United States. It is published in French and English and "enables a zoologist to determine the valid name for a taxon to which an animal belongs at any rank in the hierarchy subspecies, species, genus, and family." The *Code* includes a glossary, index of scientific names, and "Tables and explanatory notes, compiled as an aid to zoologists" (pp. 203-229).

Parker, Sybil P. *Synopsis and classification of living organisms.* New York: McGraw-Hill; 1982. 2 vols.
A thorough, detailed, and scholarly presentation of classifications for all living organisms.

Preston-Mafham, Rod; Preston-Mafham, Ken. *Spiders of the world.* New York: Facts on File Publications; 1984. 191 p.
Describes and illustrates the characteristics of spiders. Has an international scope.

REFERENCES

1. Cowan, S. T.; Hill, L. R., eds. *A dictionary of microbial taxonomy.* Cambridge: Cambridge University Press; 1978: p. 257.
Designed to aid individuals who find themselves needing to understand the language of taxonomy as applied to microbiology, including terms used in botany and zoology. It includes chapters discussing source materials for taxonomy, the philosophy of classification, and the early history of bacterial classification.

2. Daily, Howell V.; Linsley, E. Gorton. Taxonomy. In: Gray, Peter, ed. *The encyclopedia of the biological sciences.* 2d ed. New York: Van Nostrand Reinhold; 1970; p. 920.
This article is one of over 800 contained in the *Encyclopedia.* It aims at providing succinct and accurate information for biologists in fields in which they are not themselves experts as well as a comprehensive source for others.

3. Blackwelder, Richard E. *Taxonomy: a text and reference book.* New York: Wiley; 1967: p. 3.

The author states that this "is a book *about* taxonomy *for* taxonomists." It is arranged in six parts: Introduction (with definitions and importance); Practical taxonomy; Diversity to be classified; Classification, naming, description; Theoretical taxonomy; and Zoological nomenclature.

4. Daily, p. 920.

5. Daily, p. 924.

Chapter 31
Textbooks

HIGHLIGHTS OF TEXTBOOKS

Description: Textbooks, as their name implies, are books written primarily for classroom use, normally serving as the works around which educational courses are planned. Although they have widely different levels of difficulty, depending upon the educational use for which they are intended, they have common features which mark them as a separate category from monographs.

Significance: Textbooks play an important role in education, primarily serving as a basis for learning. They might range from books used in introductory science courses in grade schools to texts suitable for graduate education in universities. Their vocabularies must be carefully chosen to match the expected level of their readers; problems for class use are often found at the end of chapters. Unlike monographs, they usually present little or nothing in the way of unique material.

Quantity: Thousands of textbooks are published each year. Textbooks require periodic revision to keep them current and to meet changing educational standards.

Physical Characteristics: Textbooks vary in appearance largely because their intended audiences differ greatly. Styles of typeface, use of artwork, even types of covers used, are chosen with the educational level of the targeted audience very much in mind. A textbook for a college course in organic chemistry would not have much in common in appearance with a chemistry text for high school students, yet both are designed for use in classroom instruction.

Availability: Many libraries do not collect textbooks extensively since texts, by virtue of their very nature, do not usually contain original material; the contents of most texts have already been published in some other format, such as a monograph or a periodical article. Textbooks may thus not even be included in a library for a particular school unless copies are kept on reserve.

Retrieval: Textbooks are no different from monographs in regard to retrieval; both may be located by author, title, subject, and series in the usual library catalogs or in special finding tools, which might be printed or computerized.

Intended Audience: As previously discussed, textbook audiences may range from grade school children to doctoral students in universities. Lay people may find textbooks useful, depending upon their level of difficulty.

Scope: Textbooks tend to cover much wider topics than monographs. Having a wider scope, textbooks must of necessity treat topics in less detail than monographs. Textbooks can present an overview of a subject, while monographs aim at full coverage of a more narrow topic.

NATURE OF TEXTBOOKS

Textbooks are usually created for use in classes, thus requiring that the author(s) use a tutorial style. Sample problems are commonly included, as are chapter summaries of major points for class discussion. Monographs would not have either sample questions or discussion points. It should be noted, however, that monographs are sometimes used in graduate instruction, where students are advanced in the subject to the point of being able to understand a more advanced treatment. Such monographs contain few, if any, of the features commonly found in textbooks.

The steady development of advances in sci-tech disciplines requires that textbooks be updated every few years (often every four to five years) in order to keep up with the times. On the other hand, monographs might conceivably be written in a more philosophical or theoretical fashion and thus would not be affected by the changes in techniques or applications that require revisions of textbooks. Sometimes it is hard to understand why new editions of certain textbooks are published. For example, it is difficult to imagine great changes in a branch of basic mathematics, say calculus, that would necessitate the publishing of a new edition every four years, yet this does happen.

One of the best ways of locating data about both monographs and textbooks is by using a well-known indexing tool for books in these disciplines, *Scientific and Technical Books and Serials in Print.* This annual listing provides information about the authors, titles, subjects, prices, and publishers of more than 100,000 sci-tech books, as well as data on thousands of sci-tech periodicals.

Some textbooks are challenged by community standards and customs when certain topics are involved, such as texts dealing with evolution, abortion, and birth control. In some states the censorship of texts, particularly for primary and secondary grades, presents authors and publishers with difficult decisions. Frequently the only way authors can have their books accepted as textbooks in certain communities or certain states is to write them in such a fashion as to meet local standards or regulations, whether appropriate to the content of the books or not. Court cases have resulted in some instances when publishers wished to seek legal decisions to support their right to use their best judgment regarding the subjects and positions taken in their textbooks. A few publishers refuse to change their texts to meet local standards, even if that means loss of sales in some communities. The issue of community control over local schools versus the judgment of professional educators and publishers is long standing in this country. This problem rarely affects college-level textbooks, except in a few cases, such as books for schools sponsored by particular religious sects.

Since textbooks rarely, if ever, contain previously unpublished data of a research nature, most of them can be classified as secondary sources. Monographs, on the other hand, quite often present information never before published. Such monographs would thus be primary sources. For example, a monograph might be totally dedicated to the principles of designing microchips for computers, including data that the author(s) may have discovered in the course of doing research on the subject. Perhaps the book represented the first instance in which the concepts were fully covered; previously, the authors may have written a periodical article that gave only a bare outline of the major points, preferring to save full details for a monograph. By contrast, a textbook on the subject might aim at helping the student understand very basic points about the operation and purpose of microchips rather than giving them detailed information on the subject.

Because of the more or less routine, basic level of textbooks, most sci-tech libraries would do well to keep the proportion of textbooks in their collections quite low, since such books add little or nothing to the advancement or developments in the disciplines they cover. An occasional textbook might be acceptable, aimed at helping a beginner in a field who wants a clear, understandable description of some topic, but it would be unwise to stock many such books.

Both monographs and textbooks are indispensable, but it is important to realize their different purposes and values in making selections for the collection or in recommending sources for library users to examine.

One important segment of sci-tech text publishing consists of books written for children. Some publishers specialize in this field, while many others include such books in their annual crop of new books. Besides the obvious need to make children's books understandable, most publishers go further and try to make them appealing and attractive. Colorful illustrations and large print are standard features of these books. Unlike sci-tech books for adults, there are few sources available to enable one to identify books for children in these disciplines. One of the most outstanding examples of this sort of tool is entitled *Appraisal: Science books for young people*. A similar publication, *Science books and films*, evaluates not only books for general audiences but also includes films and other types of audio-visual aids.

TYPICAL EXAMPLE OF A TEXTBOOK

Wallace, Robert A.; King, Jack L.; Sanders, Gerald P. *Biology: The science of life*. Glenview, IL: Scott, Foresman; 1986. 1,217 p.
Exemplifies the typical college-level textbook, including such traditional features as review questions for the reader at the end of a chapter, use of italics to highlight important terms, and summaries of the highlights at the end of chapters.

TYPICAL EXAMPLES OF TEXTBOOK SOURCES

Appraisal: Science Books for Young People. Boston: Children's Science Book Review Committee; 1974– . Published three times a year.
Published by a nonprofit organization sponsored by the Department of Science and Mathematics Education of Boston University and the New England Roundtable on Children's Literature. Each book is reviewed by a specialist on children's literature and by a librarian. Shows recommended age levels.

Science Books & Films. Washington, DC: American Association for the Advancement of Science; 1965– . Published five times per year.
Said to be the only publication that provides expert reviews of current books and nonprint materials in all the sciences and for all ages. Each issue contains evaluations of some 300 books, films, videocassettes, and filmstrips for general audiences, teachers, and students.

Scientific and Technical Books and Serials in Print. New York: Bowker; 1971– . Annual.
Lists around 120,000 titles of books, plus about 18,000 serials. An excellent source for checking entries, including prices. Has author, title, and subject indexes. It is more than a spin-off of the main *Books in Print* series because of its data on serials.

Chapter 32
Thesauri

HIGHLIGHTS OF THESAURI

Description: A thesaurus consists of lists of terms whose relationships to other terms are shown in detail. Most thesauri indicate, for the listed terms, which other terms are broader in scope, which ones are narrower in scope, and which ones are somehow related in meaning to the terms in the list.

Significance: Thesauri have come into great prominence since the creation of databases in recent years. They enable indexers who are assigning indexing terms to records being added to a database to be consistent and accurate in selecting the best terms for each record. They also make it possible for online searchers to determine which terms have been used to index documents in a particular online database. In a sense a thesaurus serves the same purpose as a list of subject headings used for a given catalog.

Quantity: There are several score of thesauri currently in use. Many have been created in the past 10 years or so for the reasons stated above. Not every database has one, but the ones that do are usually more accurately and quickly searched than those that do not.

Physical Characteristics: In appearance thesauri resemble dictionaries except they generally do not include definitions of terms; they usually just indicate relationships between terms.

Availability: Most thesauri are available from the producers of printed indexing services or owners of online databases. Libraries engaged in online searching often have some sort of thesauri in their collections.

Retrieval: Many catalogs use "Thesauri" as a subject heading, as well as a sub-head under the name of a topic. In addition many online database catalogs mention the name of printed thesauri which are used in the preparation of indexing for the databases.

Intended Audience: There are two main groups of thesauri users. One group is made up of indexers who use the thesauri in analyzing and indexing records going into a database or a printed indexing service. The other group consists of searchers of online databases or printed indexing services who use thesauri in order to do more efficient and effective searching.

Scope: Thesauri are usually confined to a particular discipline or set of related disciplines. One very well-known thesaurus covers all areas of psychology, while others might be restricted to smaller fields, such as nonferrous metals. Thus the size and scope of thesauri vary greatly.

NATURE OF THESAURI

A thesaurus is a list of terms which may or may not include their meanings but invariably indicates their relationship to other terms. For example, a thesaurus listing of "aircraft" would show that it is a narrower term than "air transportation" but is a broader term than "bombers" or "fighter planes" or "commercial aircraft." The entry would probably also show that the term "aircraft" was related to such terms as "balloons" or "lighter-than-air craft" or "blimps." It might also show that the term "aircraft" is preferred to "airplane." In other words, the reader could easily find related terms to search in a database or could learn that a term originally picked for a search might not be as accurate or fruitful a choice to use as would an alternate term found in the thesaurus entry.

Over the years, the structure of thesauri has changed. Many of those published in the late 1950s and 1960s use very different methods for showing relationships between terms and concepts. Those published in recent years tend to use consistent structures because of the standard for thesaurus construction developed by the American National Standards Institute's National Information Standards Organization.[1] This consistency in presentation of relationships enables anyone familiar with the standard form to understand the thesaurus, even if they do not understand the field of study they are attempting to search.

Most thesauri are centered on rather specific disciplines, such as metals, but a few cover many disciplines, such as all of engineering or all of biology. Many are issued by organizations responsible for preparing databases. Searchers generally rely on such thesauri when they are available, readily buying new editions in order to keep their search techniques up to date. Figure 32-1 is a sample page from a well-known thesaurus, *Thesaurus of Metallurgical Terms*. Note that the year a term was added to the list is indicated in parentheses after the term, one sign of a carefully maintained thesaurus.

Figure 32-1. *Thesaurus of Metallurgical Terms*—Sample Page

Tile (material)

Tile (material) (1966)
  RT Ceramics
      Soil pipe

Tilt (1990)
  USE Camber

Tilting furnaces (1966)
  HS Furnaces
      . Batch type furnaces
      .. Tilting furnaces
      ... Tilting openhearth
          furnaces
  RT Electric furnaces
      Gas fired furnaces
      Oil fired furnaces
      Open flame furnaces
      Rotary furnaces
      Steel making

Tilting openhearth furnaces (1966)
  HS Furnaces
      . Batch type furnaces
      .. Tilting furnaces
      ... **Tilting openhearth
          furnaces**
      . Open flame furnaces
      .. Reverberatory furnaces
      ... Openhearth furnaces
      .... **Tilting openhearth
           furnaces**
  RT Ajax furnaces
      Basic openhearth process
      Duplex openhearth process
  UF Tandem furnace process

Tilting rotary furnaces
  USE Rotary converters

Timber (structural) (1966)
  HS Materials
      . Construction materials
      .. Structural materials
      ... **Timber (structural)**
      . Wood products
      .. **Timber (structural)**
  RT Wood
  UF Structural timbers

Time (1966)
  HS **Time**
      . Dwell time
      . Lifetime
      . Response time
  RT Astronomical instruments
      Exposure
      Frequencies
      Synchronism

Time measurements (1966)
  HS Measurement
      . **Time measurements**
  RT Cathode ray oscilloscopes
      Clocks
      Motion pictures
      Oscillographs
      Radioactive age determination
      Synchronism
      Timing devices
      Velocity measurement
  UF Timing

Time quenching
  USE Interrupted quenching

Time temperature transformation
curves (1990)
  USE TTT curves

Timers
  USE Timing devices

Timing
  USE Time measurements

Timing devices (1966)
  HS End uses
      . Machinery and equipment
      .. Measuring instruments
      ... **Timing devices**
      .... Clocks
  UF Intervalometers
      Timers

Tin (1966)
      (Chemical symbol Sn)
  HS Materials
      . Metals
      .. Nonferrous metals
      ... Heavy metals
      .... **Tin**

Tin base alloys (1966)
      (Alloys Index code SN)
  HS Materials
      . Metals
      .. Alloys
      ... Nonferrous alloys
      .... Heavy metal alloys
      ..... **Tin base alloys**
  RT Fusible alloys
      White metal
  RF Antifriction alloys

Tin bronzes (1966)
      (Bronzes in which Tin is the main
       alloying element)
  HS Materials
      . Metals
      .. Alloys
      ... Transition metal alloys
      .... Copper base alloys
      ..... Bronzes
      ...... **Tin bronzes**
      .... Leaded bronzes
      .... Phosphor bronzes
  RT Aluminum bronzes
      Beryllium bronzes
      Cadmium bronzes
      Manganese bronzes
      Silicon bronzes

Tin compounds (1966)
  HS Compounds
      . **Tin compounds**
      .. Stannates
  RT Intermetallics

Tin free steel (1990)
      (A product having a multilayered
       coating and used as a
       substitute for tin plate)
  HS Materials
      . Metals
      .. Alloys
      ... Transition metal alloys
      .... Ferrous alloys
      ..... Steels
      ...... **Tin free steel**
  RT Chromium plating
      Galvanized steels
      Precoated strip
      Tin plate

Tin nickel alloy plating
  SEE Alloy plating
       Nickel plating
       Tin plate

Tin ores (1966)
  HS Ores
      . **Tin ores**
      .. Cassiterite

Tin plate (1966)
  HS Materials
      . Metals
      .. Alloys
      ... Transition metal alloys
      .... Ferrous alloys
      ..... Steels
      ...... **Tin plate**
  RT Detinning
      Tin free steel
  RF Tin nickel alloy plating

Tin plating (1966)
  HS Deposition
      . Coating
      .. Plating
      ... **Tin plating**
      . Surface finishing
      .. Coating
      ... Plating
      .... **Tin plating**
  UF Electrotinning

Hot dip tinning
Hot tinning
Tinning (coating)

Tinning (coating)
  USE Tin plating

Tinning (soldering)
  USE Soldering

Tires (1966)
  RT Landing gear
      Wheels
  UF Tyres

Titanates (1966)
  HS Compounds
      . Transition metal compounds
      .. Titanium compounds
      ... **Titanates**
  RT Ceramics

Titanium (1966)
      (Chemical symbol Ti)
  HS Materials
      . Metals
      .. Nonferrous metals
      ... Light metals
      .... **Titanium**
      . Transition metals
      .. **Titanium**

Titanium base alloys (1966)
      (Alloys Index code TI)
  HS Materials
      . Metals
      .. Alloys
      ... Nonferrous alloys
      .... Light metal alloys
      ..... **Titanium base alloys**
      ... Transition metal alloys
      .... **Titanium base alloys**

Titanium carbide (1966)
  HS Compounds
      . Chemical compounds
      .. Inorganic compounds
      ... Ceramics
      .... **Titanium carbide**
      . Metalloid compounds
      .. Carbon compounds
      ... Carbides
      .... Metal carbides
      ..... **Titanium carbide**
      . Transition metal compounds
      .. Titanium compounds
      ... **Titanium carbide**
      Materials
      . Ceramics
      .. **Titanium carbide**
  RT Cutting tool materials

Titanium compounds (1966)
  HS Compounds
      . Transition metal compounds
      .. **Titanium compounds**
      ... Titanates
      ... Titanium carbide
      ... Titanium dioxide
      ... Titanium nitride
  RT Intermetallics

Titanium dioxide (1966)
  HS Compounds
      . Chemical compounds
      .. Oxygen compounds
      ... Oxides
      .... Dioxides
      ..... **Titanium dioxide**
      . Transition metal compounds
      .. Titanium compounds
      ... **Titanium dioxide**
  RT Pigments

Titanium nitride (1982)
  HS Compounds
      . Chemical compounds
      .. Inorganic compounds
      ... Ceramics
      .... **Titanium nitride**
      . Nitrogen compounds
      .. Nitrides
      ... **Titanium nitride**
      . Transition metal compounds

      .. Titanium compounds
      ... **Titanium nitride**
      Materials
      . Ceramics
      .. **Titanium nitride**

Titanium ores (1966)
  HS Ores
      . **Titanium ores**
      .. Ilmenite
      .. Rutile

Titanium plating (1966)
  HS Deposition
      . Coating
      .. Plating
      ... **Titanium plating**
      Finishing
      . Surface finishing
      .. Coating
      ... Plating
      .... **Titanium plating**

Titanium steels (1966)
      (Alloys Index code SATI)
  HS Materials
      . Metals
      .. Alloys
      ... Transition metal alloys
      .... Ferrous alloys
      ..... Steels
      ...... Alloy steels
      .... **Titanium steels**
  RT High alloy steels
      Low alloy steels

Titration
  USE Volumetric analysis

TNT
  RT Explosives
  UF Trinitrotoluene

Togo (1990)
  HS Geographical locations
      . Africa
      .. **Togo**

Tokamak devices (1979)
  HS End uses
      . Machinery and equipment
      .. Reactors
      ... Nuclear reactors
      .... Nuclear fusion reactors
      ..... **Tokamak devices**
  RT Nuclear fusion
      Thermonuclear power
      generation

Tolerances (1966)
  HS **Tolerances**
      . Dimensional tolerances
  RT Acceptability
      Accuracy
      Clearances
      Drift
      Errors
      Hysteresis
      Inspection
      Materials testing
      Measurement
      Nondestructive testing
      Precision
      Quality control
      Reliability
      Reproducibility
      Resolution
      Stability

Toluene (1966)
  HS Compounds
      . Chemical compounds
      .. Hydrogen compounds
      ... Hydrocarbons
      .... **Toluene**
      .. Organic compounds
      ... Aromatic compounds
      .... **Toluene**
      .. Hydrocarbons
      .... **Toluene**
      . Metalloid compounds
      .. Carbon compounds
      ... Hydrocarbons
      .... **Toluene**

228

The need for periodic updating of these lists is obvious, especially in science and technology. The versions of thesauri used by indexers responsible for analyzing the records being added to databases or printed indexes are updated more or less continuously, even

---

Thesaurus of Metallurgical Terms, 9th edition, 1990. Reprinted with permission of Materials Information, ASM International.

if the printed editions for the general public are not revised frequently.

TYPICAL EXAMPLES OF THESAURI

Medical Subject Headings. Bethesda, MD: National Library of Medicine. Annual.
MeSH is a supplement to the January issue of *Index Medicus.* It is a primary listing of subject descriptors used in the index, presented in two ways. The "Alphabetic List" provides an alphabetical arrangement of all subject headings and cross-references. The "Tree Structures" arranges the subject headings into categories and subcategories, with alphanumeric designations; this provides a classified approach to terms, as distinct from an alphabetical one. This publication provides users of *Index Medicus* with the terms to use in searching each year's volumes of the printed index.

Online searchers have a similar tool, known as the *Annotated Medical Subject Headings*, published annually by the National Library of Medicine (NLM). This publication contains headings and other access terms not provided in the print version of *Index Medicus*, but that are searchable online. It includes indexing and cataloging annotations, history of the use of the terms, and online notes, along with appropriate uses of sub-headings.

Thesaurus of metallurgical terms. 9th ed. Materials Park, OH; ASM International; 1990. 259 p.
Consists of 6,640 terms used in *Metals Abstracts Index*, as well as in METADEX, the online version of the printed index. Serves as a vital tool for searching data on metallurgy, including alloys. The *Thesaurus* also contains terms used in the metals-related portion of the *Materials Business File.*

REFERENCE

1. National Information Standards Organization. *Guidelines for thesaurus structure, construction, and use.* Z39.19-1980. New York: American National Standards Institute; 1980. 20 p.
Provides guidance in the preparation and use of thesauri; one purpose is to promote consistency among the various thesauri being created.

Chapter 33
Translations

HIGHLIGHTS OF TRANSLATIONS

Description: Translations are publications that have been transformed from one language to another.

Significance: Science and technology have no barriers when it comes to languages. Time has shown that it is unwise to ignore publications just because they are written in foreign languages. Translations help make it possible for people who have no understanding of foreign languages to grasp the meaning of publications from other countries.

Quantity: There are hundreds of thousands of translations devoted to science or technology. Most are translations of periodical articles, followed, in order of quantity, by patents, technical reports, and books.

Physical Characteristics: Translations have no distinctive appearance, usually appearing in the same form as the original publications, such as serials or monographs.

Availability: Translations can sometimes be purchased inexpensively from certain government agencies. Another method is to purchase or subscribe to cover-to-cover translations of certain foreign-language journals. The most expensive sources are commercial translating agencies which provide translations for a fee. In some countries, jointly sponsored translation centers have been established to provide documents at a modest cost.

Retrieval: Translations are indexed in various ways, including a few services devoted exclusively to them. Other indexing services include translations but make no specific effort to treat them differently from original-language materials.

Intended Audience: The users of translations are generally professionals who are experienced in their fields; the layperson has little need for such materials.

Scope: Translations have the same scope as the material being translated, including journal articles, books, patents, and technical reports.

NATURE OF TRANSLATIONS

In view of the worldwide devotion of time and money to science and technology, it would be foolish for scientists and engineers in any country to ignore the contributions of their colleagues in other countries. One example of this can be seen in the surprise experienced by many American engineers when the Soviets were the first to launch a successful space satellite in 1957, four months before the first United States satellite. Had aeronautical and astronautical engineers and top administrators been keeping abreast of Soviet accomplishments in various fields of technology, they might not have been so surprised by Soviet accomplishments at that time. One of the chief ways by which developments of other countries can be monitored is by using translations of foreign literature.

For many years translations have been part of the rich body of information available to scientists and engineers. They exist in several formats, such as books, journal articles, and technical reports. They have been published by commercial publishers, government agencies, and professional societies. Although there are a half dozen or so languages which are translated quite often, there are few languages which have not been the subject of sci-tech interest (and subsequent translation).

One of the oldest aids to translators is the bilingual or multilingual dictionary, previously discussed in Chapter 16. These dictionaries are chiefly of use to persons making a translation of a piece of literature, rather than a tool for the casual reader.

One by-product of the concern raised in the United States over early Soviet aerospace successes was the stimulation given to the publishing of cover-to-cover translations of important Soviet sci-tech journals. Although completion of these translated versions often did not occur until months after publication of the original edition, the cost of a subscription (often several hundred dollars a year) was still much less than the cost of having only a few articles translated by professional translating firms. Many of the first cover-to-cover translations were subsidized by government agencies, which were able to bear much of the cost. Other early sponsors of such translated periodicals were certain professional organizations made up of scientists and engineers. In subsequent years, cover-to-cover translations became available for journals published in other languages, including German and Japanese.

Some journals published in non-English languages frequently provide abstracts of the contents of the articles, translated into English and certain European languages, such as French and German.

It should be noted that indexing and abstracting services published in the United States have, from time to time, prepared summaries of the percentages of literature appearing in different languages. A summary prepared in 1967 shows that English was the most popular language, followed by Russian, then by German and French.[1] By 1980, a survey made by *Chemical Abstracts* showed English at 63% of the articles indexed, Russian at 20%, German at 5% and Japanese at 4.7%, followed by French at 2.4%.[2] The surge of Japanese into fourth place should not be surprising in view of the importance given to research and development in that country.

For many years an important source of translated sci-tech literature in this country was the National Translation Center, located at the John Crerar Library of the University of Chicago until 1989, when it was moved to Washington to be operated by the Library of Congress. The Center had grown from a collection of some 1,500 translations donated in 1953 by the Special Libraries Association's Translation Pool to an imposing total of over 325,000 translations, due to contributions by individual libraries as well as through aid from federal contracts. Its services and collections at Crerar were described in a paper by Ildiko Nowak.[3] In the same periodical issue there are articles about the sources of translations and the work of translators, emphasizing their use in sci-tech libraries.[4] The National Translation Center published various lists and translation indexes (including the *Translations Register-Index*) which listed over 300,000 citations to translations accumulated.[5]

Another important source of translations is the International Translations Centre, formed in 1960 as a cooperative service sponsored by over a dozen western European countries. Its announcement bulletin is known as *World Translation Index*, published since 1987 and continuing two earlier translation indexes.[6] It is published in cooperation with the Library of Congress and the National Technical Information Service (NTIS).

Several abstracting and indexing services (and their online equivalents) frequently include translations, as exemplified by those published by the National Aeronautics and Space Administration (NASA) and NTIS.

Many companies involved in sci-tech research have full-time translators on the staffs of their libraries or information centers. The familiarity with company projects which these translators possess makes their translations doubly valuable since they are aware of the precise terminology to use in translating foreign literature on topics relating to company interests. Most staff translators in organizations are also available for quick translations of titles, key phrases in articles, or abstracts of articles. In some organizations the library staff has an index of employees who have linguistic skills and can provide occasional brief translations upon request.

Organizations lacking staff translators, or individuals and firms seeking translating service (presumably unable to locate translations in the centers previously named), generally rely on commercial translating companies. Besides having skilled full-time translators on their staffs, these companies usually maintain a large roster of part-time translators who collectively cover many languages and subject backgrounds, being particularly valuable for languages that are less common. In 1988 a typical translating company charged $7.00 per 100 words and up, with higher charges made for the less frequently used languages or for more technical data.

REFERENCES

1. Wood, D. N. The foreign language problem facing scientists and technologists in the U.K.: Report of a recent survey. *Journal of Documentation*. 23(2): 117-130; 1967 June.
Analyzes the relative frequency of different foreign languages as found in the articles covered in six major English-language indexing and abstracting services.

2. CAS today: facts and figures about *Chemical Abstracts Service*. Columbus, OH: Chemical Abstracts Service; 1980: p. 8.
Presents an analysis of languages frequency found in the papers indexed by this publication.

3. Nowak, Ildiko D. The National Translation Center: its development, scope of operation and plans for the future. *Science & Technology Libraries*. 3(2): 13-19; 1982 Winter.
Reviews the history of the Center, its services, and its holdings.

4. Mount, Ellis, ed. *Role of translations in sci-tech libraries*. New York: Haworth Press; 1982. 94p. (Also published as *Science & Technology Libraries*, v.3 no. 2, 1982 Winter)
Contains papers about the value of translations, use of commercial translators, commercial translating firms, the role of the free-lance translator, and similar papers.

5. *Translation Register Index*. Chicago: National Translation Center; 1967-1986. Semi-monthly.
An index to hundreds of thousands of translations in the Center until 1989, when it was moved to the Library of Congress. Provides an index by patent number and by journal citation, giving a full bibliographic citation for each item. Continues various lists and indexes that began in 1959.

6. *World Translation Index*. Delft, Netherlands: International Translation Centre; 1987– . Monthly.
Lists translations available at the Centre. Includes translations of periodical articles, patents, and standards. Adds around 30,000 titles annually. Has a source index as well as an author index. Continues the *World Transindex* (1959-1986) and *Translation Register Index*.

PART IV

Secondary Sources
of Information
(Nontextual)

Chapter 34
Audio-Visual Materials

HIGHLIGHTS OF AUDIO-VISUAL MATERIALS

Description: Audio-visual materials are familiar to all people because of their wide-spread use in society. As a class they include motion pictures, slides, phonograph records, audiotapes, videotapes, and related materials.

Significance: Audio-visual materials are generally recognized as possessing several advantages over printed materials, such as having a greater impact on viewers or listeners in certain cases, providing greater clarity in showing physical relationships among objects, and having the capability for storing visual or auditory information that cannot be retained permanently without such media.

Quantity: There are millions of pieces of audio-visual materials important to science and technology that are currently in existence, including photographs, slides, reels of motion picture film, and audio/video tapes.

Physical Characteristics: Audio-visual materials dealing with sci-tech subjects have the same appearances and features as all other audio-visual aids. The range is great, including formats involving paper, magnetized and optical materials, films, and other types of graphic products.

Availability: Someone seeking a particular audio-visual product will undoubtedly have a more difficult time locating it than if a printed product were needed. One reason is the difficulty of learning what audio-visual products exist as well as the additional problem of finding a source for it once it is known to exist. Still another problem is that of needing the proper equipment—a projector, a player or other device—for displaying or using the particular piece.

Retrieval: There is no single source for identifying and/or locating audio-visual materials, whether they involve subjects of a sci-tech nature or not. One index may cover motion pictures/filmstrips, another may deal with audiotapes/records, and still another with video-tapes. Compared to the excellent indexing services for books and journal articles, bibliographic control of audio-visual materials is poor. A few databases include them along with the usual book and journal coverage, but the coverage is spotty and unpredictable.

Intended Audience: There are audio-visual products for all ages and backgrounds, ranging from videotapes about science aimed at grade school children to slides of complicated laboratory results that only trained experts in a particular field could understand.

Scope: Audio-visual materials have a virtually unlimited scope since they can apply to any field in science and technology, can be prepared for a wide range of audiences, and can involve practically any medium used singly or in conjunction with other media.

NATURE OF AUDIO-VISUAL MATERIALS

A survey of information sources devoted to science and technology would be incomplete without a discussion of what are loosely called audio-visual aids. Most people have grown up in a society in which they have been exposed since early childhood to such media as motion pictures, slides, filmstrips, and, in the past two decades, videotapes. Educators at many levels in the school systems have generally made use of such formats, subject to budgetary limits. A picture of a certain object may make its features much more easily understood than a printed description of it. The sound of a scientist's voice giving instructions or providing leadership to a group reveals a great deal about the type of person he or she is, often more than a printed biography would provide. Modern day use of optical disks to store machine-readable data have proved their worth in computer systems.

Probably more has been done with sci-tech audio-visual materials designed for elementary and high school education than for more advanced educational levels as far as sheer numbers of available products are concerned. Nevertheless there are many instances in which even graduate-level courses rely on motion pictures and similar media.

For example, a college chemistry professor might videotape a complicated, time-consuming experiment in advance of a scheduled lecture session for hundreds of students so that they would see a successful performance, rather than risk the embarrassing (and time-wasting) experience of a failed effort. Medical and dental schools are also likely to use videotapes of delicate operations for instructional purposes; reading about such processes would be a poor substitute for watching them being performed. In such cases, however, printed descriptions would be useful supplements for pointing out details that should be stressed, or for presenting background information concerning the operations.

Videotapes are easier to create than motion pictures, particularly when budgets are low and professional camera operators familiar with making motion pictures are not available. No doubt higher quality films can be made with motion pictures than with a video process, but not every use requires such a level of quality.

Slides for projection have long been used in education, including sci-tech applications. They are relatively inexpensive and can be easily projected. A common product is a combination of slides with an audiotape that contains an explanation of the slides. These are also inexpensive and simple to produce.

Besides formal classroom use, there are several other applications of audio-visual aids. One is for promotional or marketing purposes, used by commercial manufacturers and laboratories to inform potential customers of the features of their products. Utility companies and telephone companies, having a need to ensure good relations with thousands of customers, know the goodwill and educational value of high-quality motion pictures, aimed for use by clubs, schools, and miscellaneous groups. One example of this application is a color motion picture showing how microchips are manufactured; this has served a commercial laboratory wanting to inform the public about its skills and fields of activity. Other examples include videotapes made by organizations to inform employees about the workings of inhouse departments, such as libraries and information centers.

Still another use is by publishers, who create such products as tape/slide presentations to inform users about the features of their more complicated products. An example is the audio-visual product prepared to explain the use of *Beilstein's Handbuch der Organischen Chemie*, probably the most complicated reference tool yet created. The tape/slide show makes this index of organic chemicals less difficult to comprehend.

For similar reasons, database vendors have created video tapes for use by individuals needing help in mastering the steps in doing online computerized searches. Not everyone is able to enroll in formal courses for learning such skills.

One fast-moving area of development involves laser disks for storing data formerly found only on computerized databases. These disks, now usually called CD-ROMs (standing for Compact Disks—Read Only Memory) can be searched as often as desired by purchasers without the usual fees for online searching. Some CD-ROMs cannot be searched in as sophisticated a fashion as the corresponding online databases, and they cannot be as current in contents as the databases (since the updating of CD-ROMs is often no more frequent than quarterly). They are discussed in greater length in Chapter 15, "Computerized Information Sources."

In recent years combinations of online databases with visual images has attracted more interest. An example involves patents; more than one organization has developed an online database for searching the text of patents, combined with stored optical images of the illustrations that accompany the original printed patents. The illustrations might be on a videodisc or perhaps on a CD-ROM. Having both text and illustrations available at the same time is a great convenience for patent searching.

One of the limitations of audio-visual aids is the difficulty of learning what is available. With so many organizations creating such products within the confines of their own groups, having little desire to publicize them to outsiders, a great deal of duplicate effort no doubt takes place. The best-indexed audio-visual aids are those prepared for elementary and high school courses; many sales catalogs exist for such products, providing descriptions and cost information. Products that are more complicated in nature are harder for the average person to locate or even become aware of. Many periodicals include a/v materials in their reviews, but those of a sci-tech nature are in the minority. A notable example of a source devoted to reviews of recent sci-tech films (and books) is *Science Books & Films.*

Perhaps in time there will be more attention paid to audio-visual materials available to several levels of the scientific and technical communities. At this time there is a lot of experimentation by the producers of such materials to determine which ones are economically viable in the marketplace. Some products will be abandoned when sales are disappointing, while other trial products will take their place. It is a very fluid situation. The range of intended audiences is quite wide. For example, a videodisc entitled *Cell Biology* is aimed at college students, probably those majoring in biology. It contains several score of film segments on various aspects of the topic. On the other hand, there would probably be a large audience of laypersons interested in *A Field Guide to Bird Songs of Eastern and Central North America.* It consists of a set of phonograph records containing several hundred songs.

TYPICAL EXAMPLES OF SOURCES OF AUDIO-VISUAL MATERIAL

Science Books & Films. Washington, DC: American Association for the Advancement of Science; 1965– . Published five times per year.
An unusual source featuring both films and books. Consists of reviews of current books and audio-visual materials, intended for both adults and young people.

TYPICAL EXAMPLES OF AUDIO-VISUAL MATERIALS

Cell biology. Videodisc. Burlington, NC: Carolina Biological Supply; 1989.

This videodisc includes 86 film segments and additional still frames in five major categories: cell types, cell constituents, mitosis and cytokinesis, fission, and cell motility. Videodisc and image directory in English or Spanish.

Peterson, Roger Tory. *A field guide to bird songs of Eastern and Central North America.* Boston: Houghton Mifflin; 1959.

A set of phonograph records offering narration, bird songs, and bird calls that is designed as an accompaniment to Peterson's *A Field Guide to Birds* (Peterson Field Guide series; no. 1). Over 300 bird calls and songs are represented on these records.

Chapter 35
Maps and Atlases

HIGHLIGHTS OF MAPS AND ATLASES

Description: Maps and atlases are familiar to all people, but the many types which involve science and technology may not be well known to the reader. Their subject matter may consist of such varied topics as charts of the ocean floor, maps of crops in Africa, charts of the position of stars, or illustrations of human anatomy.

Significance: In many instances information must be presented in graphic format to be meaningful. A graph or a table or a description in text format are no substitutes for a map or an atlas when the relationships of the size and location of data are in question. Maps and atlases are important in most areas of science and technology, although more significant in some disciplines than in others.

Quantity: There are millions of maps in existence, most having a relationship to some area of science or technology. The ability of satellites to produce maps as they circle the globe has alone added huge quantities of maps to those made by more traditional methods. Atlases are not as plentiful as individual maps but probably exist in the thousands.

Physical Characteristics: Maps and atlases vary tremendously in their overall size, use of color, scale at which drawn, and type of material on which drawn, ranging from lightweight paper to three-dimensional plastics.

Availability: Many universities maintain extensive map collections, as do the larger public libraries. Still larger collections are located in certain government agencies, particularly those at the federal level. Atlases tend to be most numerous in those organizations maintaining large map collections, or in departments relying on graphic data.

Retrieval: Identifying maps is a more difficult task than locating particular books or journal articles simply because no large index has been created which covers a high proportion of the maps which exist.

Atlases, on the other hand, can usually be located in catalogs and guides to the literature, either by subject or by a heading such as "Atlases" under a particular subject.

Intended Audience: Maps and atlases serve many types of users, ranging from children to practicing engineers and scientists. A particular map may not interest such a varied group of people, but maps and atlases as a class have many users.

Scope: The scope of a map may range from a small area containing a few square feet to one that covers the entire earth or vast regions in the skies. Likewise, atlases may consist of maps for portions of a city or may contain drawings depicting parts of the anatomy of living things.

NATURE OF MAPS AND ATLASES

Maps and atlases need no introduction, so familiar are they to people of all ages, including children. What may not be so familiar is the wide array of maps and atlases that pertain to different scientific or technical disciplines. An agricultural expert would learn a lot from a map made using infrared cameras in satellites for charting crops growing in Southeast Asia. A civil engineer could tell from a soil survey what the problems would be if a highway were to be built in a particular area. Geologists studying the nature of the floor of the Indian Ocean would find a map that charts the features of that part of the world an indispensable aid to research. Each of these maps would interest a particular type of researcher.

Maps have a long history, and for centuries the techniques of making maps did not change a great deal. With the advent of color printing, maps became more attractive and more useful. Novel methods of constructing maps, using inventive techniques, have been the subject of experimentation over the years. Finally, magnetic and optical storage of data from which maps can be printed have brought a major change to the process of map-making in recent years.

Keeping track of what maps exist has never been an easy task. One government agency that was created to provide information on the availability of cartographic data is the National Cartographic Information Center, located in Reston, Virginia. Since its beginning in 1974 it has amassed records of millions of maps located in other government agencies, such as the National Aeronautics and Space Administration and the Geological Survey.

Still another federal agency, the Defense Mapping Agency (a part of the Department of Defense), not only has a huge collection of maps but also maintains a depository program for more than 200 organizations in some 70 countries. Each year over 150,000 maps are distributed to participants.

In recent years a number of online databases have begun to include maps along with other formats, but to date there is no single online source devoted to maps. Some of the bibliographic utilities such as OCLC and RLIN do include maps, but not on a large scale.

Large map collections are usually maintained by university libraries, often located in an earth sciences library because of the many geological maps needed for such disciplines. Large public libraries traditionally have sizable map collections, although many of them emphasize social science and business data rather than sci-tech information.

Bookstores that specialize in maps are a fruitful source; there are dozens around the country. A good bookstore advertizes its wares regularly and will often make a search for a particular map for a customer.

A number of books and articles have been written in recent years about the importance of maps in scientific and technical fields. *Role of Maps in Sci-Tech Libraries*, edited by Ellis Mount, describes sci-tech map collections in universities, public libraries, and government agencies.[1] In addition, directories of map libraries enable one to locate likely sources of maps that might be difficult to find. The directory published by the Special Libraries Association is a valuable source of data about map collections.[2]

Atlases, which are essentially bound sets of individual maps, are very similar to traditional reference books as to their availability. The most common type of atlas exhibits geographic features of different countries, states, or even cities. Atlases can also be found with much more specific features, serving a variety of disciplines. One example is the astronomical atlas, which traditionally consists of photographic plates depicting the features of a particular planet or perhaps major stars. In another discipline the climatic atlas is important, providing maps which show the climatic features of different areas of the world (such as temperature, humidity, or amount of rainfall). One other type of atlas deals with wildlife, showing locations of specific types of animals around the world. Still another type of atlas maps the anatomical features of living things. Commonly, such atlases provide detailed diagrams of the venous systems, nervous systems, and musculoskeletal system of animals and human beings. Similar atlases can be found for botanical species.

One of the best geographical atlases was issued in 1970, the *National Atlas of the United States*. A product of the U.S. Geological Survey, this atlas contains maps covering many disciplines, such as climatology, history, and geology. An atlas emphasizing astronomy is entitled *Atlas of the Universe*. It would interest novices as well as more experienced users. One specialized anatomical atlas is entitled

The CIBA Collection of Medical Illustrations. It is known as an excellent atlas of the human body, containing many detailed drawings.

TYPICAL EXAMPLES OF ATLASES

Moore, Patrick, ed. *Atlas of the universe.* Rev. ed. Chicago: Rand McNally; 1981. 271 p.

This astronomical atlas contains a very attractive set of impressive photographs and other illustrations. It includes a glossary, a list of stellar objects, and a section for beginners.

National atlas of the United States. Washington, DC: U.S. Government Printing Office; 1970. 417 p.

A classic atlas, eight years in preparation. It includes all areas of geosciences, such as topography, geology, climatology, and water resources, along with historical and political features. It has an index of more than 41,000 place names and contains over 300 multicolored maps.

Netter, Frank H. *The CIBA Collection of Medical Illustrations.* West Caldwell, NJ: CIBA Pharmaceutical Company; 1953– . 8 vols.

Each volume, some of which have several parts, provides detailed drawings of some part of the human anatomy, including the reproductive, digestive, endocrine, and respiratory systems, as well as the heart and kidneys. These volumes were produced in 1974, but the nervous system was revised in 1983. Other volumes will also be revised. This set is recognized by medical students, nurses, and allied health professionals as one of the finest examples of atlases of the human body.

REFERENCES

1. Mount, Ellis, ed. *Role of maps in sci-tech libraries.* New York: Haworth Press; 1985. 122p. (Also published as *Science & Technology Libraries.* 5(3): 1985 Spring.)

The theme of the book concerns types of maps and their sources, including the Earth Sciences Library at Stanford University, the New York Public Library's Map Division, the National Cartographic Information Center, and the Defense Mapping Agency. One paper discusses the various media being used for storage of maps, including paper, microforms, optical discs, and magnetic memories.

2. Special Libraries Association. Geography and Map Division. *Map collections in the United States and Canada; a directory.* 4th ed. Edited by David K. Carrington and R. W. Stephenson. New York: Special Libraries Association; 1985. 188 p.

A survey of 800 libraries having map collections. Besides the description of the libraries, the names of staff members are given along with the scope of the collections. Institutions are listed alphabetically by city within a given province or state. An exhaustive directory of map sources.

Appendices

Appendices

Appendix I
Typical Reference Questions

Librarians working with sci-tech materials do not gain overnight the skills necessary for handling reference questions that are typical of the queries posed in libraries dealing with these disciplines. The greater the librarian's familiarity with the features of the information sources discussed in this book, the greater the skill with which inquiries can be answered. The purpose of this appendix is to present what the compilers feel are typical reference questions and examples of suitable sources that would fill the inquirers' needs. Although the sample questions given here do not represent actual incidents, the sources recommended would be suitable for providing the requested information.

In planning this appendix we decided that it was not enough simply to list typical questions and appropriate information sources. In actual reference service many factors need to be considered, such as the characteristics of the inquirer (experienced or inexperienced), the level of information required (basic or high level), the type of library involved, and the resources at hand for meeting the need. Thus for each sample question we have given a brief description of the inquirer as well as the library setting, followed by the general type of source recommended and the specific source proposed.

PROCEDURES FOR HANDLING REFERENCE QUESTIONS

In the actual answering of reference questions, the librarian should consider a number of factors in deciding on the best source for the inquiry. Before rushing off to choose a source, the reference librarian would be well advised to ask himself or herself such questions as:

1. What information source would be apt to *contain* this sort of information? Perhaps several sources would be suitable in some cases, such as a monograph, a journal article, and a

    patent. Realization that more than one source exists would broaden the options available.

2. Of the alternative sources, which one would best suit the inquirer in regard to the amount of *time* available for locating the source? A person with only a short time may be willing to settle for a less-than-perfect source, while another person may want to get the best source regardless of the time it would take to locate it.

3. Would the *level* at which the material was written be appropriate for the inquirer (as best one can judge the latter)? A distinguished professor is not apt to want an article from a popular journal, while a high school freshman might be pleased with such a source.

4. Would the source chosen provide *current,* up-to-date information, if that is a factor? A directory printed five years ago would be a poor choice for finding the current address for a chemical engineer in Iowa.

5. Does the *collection* contain the ideal source, and, if not, does it contain a second-best source? Few collections contain the ideal answer for every question. In some cases the reference librarian must notify the inquirer that a second-best source is all that is available in the collection at that time. Perhaps time would permit obtaining an interlibrary loan.

    Other criteria for selecting a reference source would no doubt come to mind in a reference situation, but the above points are typical of the thought process a reference librarian should experience before searching for a particular title. In the long run it saves time to decide in advance what to seek rather than to search in a haphazard fashion.

TYPICAL REFERENCE QUESTIONS

1. Monograph

Question: I'm seeking a thorough discussion of defects in semiconductors.

Inquirer: A college student majoring in physics

Setting: A medium-sized college library

Proposed Source: A modern monograph written for a person with more than a layperson's knowledge of the subject

Actual Source Recommended: Using the local catalog under the category of "Semiconductors—Defects," the following book was located:

Chikawa, J. and others, eds. *Defects and properties of semiconductors: defect engineering.* Boston: D. Reidel Publishing Co.; 1987. 261 p.

2. Journal Article

Question: I'm preparing a research paper on the effects of Alzheimer's disease on families of the victims, and I need a current discussion of the subject. It's for a graduate psychology course.

Inquirer: A college senior majoring in psychology

Setting: A university health sciences library

Proposed Source: An article in a journal concerned with medical and/or psychological topics, written at a rather high level

Actual Source Recommended: Using *Index Medicus* under the subject heading "Alzheimer's Disease—Psychology," the following article was found:

Morris, L. W. et al. The relationship between marital intimacy, perceived strain and depression in spouse caregivers in dementia sufferers. *British Journal of Medical Psychology.* 61(3): 231-236; 1988 Sept.

3. Dictionary

Question: What is the meaning of the term Richter scale? I've seen it used in newspapers when there is an earthquake.

Inquirer: A high school student

Setting: A school library

Proposed Source: A definition in a dictionary written for students of this age group

Actual Source Recommended: Perusing the reference collection in the science section, the following book was located; it contained a definition under the term "Richter scale":

McGraw-Hill dictionary of scientific and technical terms. New York: McGraw-Hill; 1984. 1,781 p.

4. Patent

Question: I'm trying to locate a patent on the purification of oil-soluble alkaline earth sulphonates.

Inquirer: A middle-aged adult who states he is a chemist

Setting: A large public library

Proposed Source: A patent

Actual Source Recommended: Using the online database WORLD PATENTS INDEX, the following patent was located:

Kekenak, J. *Oil-soluble alkaline earth sulphonate(s) purification.* Czechoslovakian patent CS7700107-A. Jan. 1, 1977. 77CS-000107; C11D-001/12.

5. Technical Report

Question: I'm looking for a report prepared in India on the subject of metal particles, created by wear, that are found in used engine oils.

Inquirer: A lubrication engineer

Setting: A government research laboratory

Proposed Source: A technical report

Actual Source Recommended: Using the 1988 subject indexes for *Government Reports Announcement and Index* under the heading of "Wear," the following report was located:

National Aeronautical Laboratory, Bangalore (India). *Analysis of wear metals in engine oils using the atomic absorption spectrophotometric method.* By M. Kalyanam and others. Bangalore, India: the Laboratory; 1987 Aug.; NAL-TM-MT-8704; N88-13435/8/GAR.

6. Encyclopedia

Question: I have 15 minutes before a meeting at which I need to know at least the rudiments of how a jet aircraft engine works.

Inquirer: The head of the purchasing department of a company that is involved in design of a number of products

Setting: A technical library in a manufacturing company

Proposed Source: A sci-tech encyclopedia

Actual Source Recommended: Using the index of the *McGraw-Hill encyclopedia of science and technology* (6th ed.), shelved in the reference section, the following article was located: "Jet propulsion," v. 9, p. 454-457.

7. Guide to the Literature

Question: I'm about to make a thorough search on what has been written on the subject of polymer chemistry before picking a topic for my dissertation. How can I get an overall view of the types of literature created in the past 15 years that I should examine?

Inquirer: A doctoral student in chemistry

Setting: A university chemistry library

Proposed Source: A guide to chemical literature

Actual Source Recommended: Using the local catalog under the heading of "Chemical Literature," the following book was located:

Maizell, Robert E. *How to find chemical information: a guide for practicing chemists, educators and students.* 2d ed. New York: Wiley; 1987. 402 p.

8. Standards

Question: I need to know the proper mix of blacktop paving material for building a parking lot.

Inquirer: Head of a small construction company

Setting: A public library in a medium-sized city

Proposed Source: A standard

Actual Source Recommended: Using the index to *ASTM Standards 1989,* shelved in the reference section, the following standard was located:

American Society for Testing & Materials. *Standard specification for coarse aggregate for bituminous paving mixtures.* Philadelphia: ASTM; 1988; ASTM D692-88. 2 p. (Found in 1989 *Annual Bulletin of ASTM Standards,* Sec. 4, vol. 04.03, p. 158-159.)

9. Conference Literature

Question: There was a conference in 1985 on the subject of biological controls of protected agricultural crops. I'm looking for a paper given there.

Inquirer: A professor involved in environmental engineering

Setting: A biology library at a university

Proposed Source: A conference paper

Actual Source Recommended: Using the Permuterm index of *Index to Scientific and Technical Proceedings* for 1985-88 under the term "Protected Crops," the following paper was located in the 1988 volume:

Meeting of the European Committee Expert Group on Integrated and Biological Control in Protected Crops. Held in Heraklion, Greece, Apr. 24-26, 1985. Published as:

Integrated and biological control in protected crops. Edited by R. Cavalloro. Brookfield, VT: Balkem Publishers; 1987. 251 p.

10. Field Guide

Question: There is a little snake that is living in our yard. What can I use to find out if it is poisonous or not?

Inquirer: A worried housewife

Setting: A small public library in the mid-Atlantic region of the United States

Proposed Source: A field guide that will identify the snake by either a photograph or color drawing and that will also provide information about the nature of the snake

Actual Source Recommended: Using the reference collection, the following source was located:

Conant, Roger. *A field guide to reptiles and amphibians of Eastern/Central North America.* Illustrated by Isabelle Hunt Conant. New York: Houghton Mifflin; 1975. 448 p.

11. Directory

Question: I'm trying to locate a woman I met at a conference held in Europe in 1987. She's a German botanist, her first name is Maria and her last name is something like Ehberg.

Inquirer: A woman who is writing a book on plants

Setting: A science library at a college

Proposed Source: A biographical directory

Actual Source Recommended: Using *Who's who in science in Europe,* 5th ed., located in the reference section, the following information was found in the chapter on German scientists, under "Botany":

Maria E. Ehrenberg. Botany Institute, Wurzburg, Germany.

12. Handbook

Question: I need a listing of some of the applications of industrial precipitators.

Inquirer: A mechanical engineer

Setting: Technical library of a manufacturing company

Proposed Source: A handbook

Actual Source Recommended: Using the reference section of the library, the following book was located:
Marks' standard handbook for mechanical engineers. 9th ed. Edited by Eugene A. Avallone and Thedore Baumeister III. New York: McGraw-Hill; 1987: p. 18-13, 18-18.

* * * * *

In conclusion, it should be noted that not every sci-tech librarian would agree on either the basic type of source recommended or the actual choice, but, nevertheless, it is likely that the sources recommended in this chapter would fit the inquirers' needs in an acceptable manner.

Appendix II
Glossary

Almanac. A publication, usually issued annually, that contains statistical, tabular, and/or general information.

Annual. Any publication issued once per year.

Anthology. A collection of writings, passages, or extracts from the works of several authors on a particular subject.

Atlas. A bound collection of maps, charts, plates, and so on, illustrating a subject. Most atlases commonly consist of maps.

Bibliography. A listing of books and other materials having a common feature, such as information on a given topic, items contained in a particular collection, or items similar in format or physical aspects. Several types of bibliographies exist, including current—consisting of newly produced materials; retrospective—consisting of older materials on a topic; recurring—published at some identifiable interval; and literature searches—in-depth lists of references, usually created on demand and often produced with the use of computerized databases.

Conference Literature. *See* Proceedings.

Database. A computerized collection of information or data. Includes bibliographic databases, which consist of listings of citations to literature, and nonbibliographic databases, which provide such data as statistics, full text of papers, and chemical formulae.

Dictionary. A listing of terminology in general usage, often in specific subject fields. Some of them provide spelling, meaning, etymology, and pronounciation. Dictionaries can be restricted to a single language, can be multilingual, or can be bilingual.

Directory. An organized listing of people, products, companies, services, events, or any combination of these.

Dissertation or Thesis. Unbiased, monitored reports of an individual's research, often containing summaries of the state-of-the-art on a topic, and generally containing good bibliographies.

Encyclopedia. A work, usually written by specialists, with articles, arranged in alphabetical order, that include history, discussions, and bibliographies.

Field Guide. A work which provides a visual approach to the study of some aspect of nature, such as geological formations or birds or flowers.

Guide to the Literature. A survey intended to suggest sources for further information. It describes the types of literature available, gives examples with annotations, often evaluates contents of works or compares them with others, and sometimes guides readers in the use of libraries or information centers.

Handbook. A compact reference source written for experts in the field, containing concise compilations of data or descriptive information, logically arranged in numerous chapters.

History. A retrospective overview of a particular subject.

Journal. A serial intended for a scholarly audience, usually issued at regular intervals, more than once per year, containing writings by several authors. *See also* Periodical; Refereed Journal; Serial.

Manual. A publication that provides rules or guidelines for processes, explains procedures, and is generally prescriptive in nature.

Manufacturers' Literature. Often known as trade catalogs, such sources identify a manufacturer's products, including descriptions, applications, and cost.

Monograph. A printed work that provides comprehensive coverage of a single topic. It is a complete bibliographic unit. It may be part of a series issued in successive parts at regular or irregular intervals.

Newsletter. A periodical usually devoted to a particular subject field that provides news items of interest to the members or subscribers, along with announcements and letters from readers.

Newspaper. A periodical issued at least once a week on newsprint paper in folded format, primarily devoted to feature articles, commercial advertisements, local announcements, and miscellaneous information.

Patent. A document establishing a government-granted right to exclusive ownership of the design of a particular product or the operation of a particular process. Written documentation includes detailed diagrams, abstracts, and carefully written descriptions of the product or process; may include a bibliography.

Periodical. Any publication usually issued at regular intervals, more than once per year, containing writings by several authors. It is usually aimed at a broader audience than a journal, which is primarily designed for a more scholarly readership. *See also* Journal; Refereed Journal; Serial.

Preprint. Any copy of a work available for informal distribution before intended publication.

Proceedings. A report containing partial or full coverage of the papers presented at a gathering, such as a conference, symposium, meeting, workshop, or colloquium, to name a few titles for such events. May appear as a separate volume, a special issue of a journal, or as separate journal articles issued over a period of time.

Pseudo-serial. A reference work that is issued in successive editions and treated as a serial, such as sets of directories, specialty handbooks, and biographical lists.

Refereed Journal. A journal in which all the papers submitted are evaluated by experts in the field prior to publication to assess quality. *See also* Journal; Periodical.

Report, Research. A document that provides an account of progress or gives the results of a project.

Report, Technical. A document that provides information about a specific project, produced by organizations most often for internal distribution, specifically for knowledgeable people.

Review of the Literature. A survey of important publications on a selected topic, sometimes focused on works published in a single year.

Serial. Any publication issued at intervals in sequentially numbered or dated parts and intended to continue indefinitely. Also, a type of publication, other than a periodical, that is typically issued either at regular intervals of one per year or less often, or at irregular intervals, no matter what the frequency. *See also* Journal; Periodical; Series.

Series. A type of serial with parts that can be treated as separate works or monographs. *See also* Serial.

Standard and Specification. A document that gives detailed instructions for approved methods of making a particular product or performing a certain process.

Table. A work that consists of data that are essentially numerical in nature, arranged in tabular form.

Technical Report. *See* Report, Technical.

Textbook. A teaching instrument, complete but not necessarily comprehensive, designed for classroom use.

Thesaurus. A list of accepted descriptors or indexing terms, often with an elaborate structure, for use in the organization of a collection of information documents.

Thesis. *See* Dissertation or Thesis.

Translation. A work originally written in one language that has been rendered into another language.

Treatise. A work that provides comprehensive historical coverage of a single topic, with complete and authoritative compilations and summaries of the subject, suggested areas for further research, and extensive documentation. It contains methodical discussions of the facts and principles involved, conclusions reached, and evaluations of known information. The work usually requires expertise in the subject to understand.

Union List. A location guide to books, serials, and other materials, usually with full bibliographic information and identification of holdings of selected libraries.

Yearbook. An annual publication that provides descriptions of events of the year covered, usually with facts or figures. Often issued to update a major reference.

Index

ELLIS MOUNT

Ellis Mount has been a member of the faculty of the School of Library Service at Columbia University since 1977. Taking partial retirement in 1989 has enabled him to spend more time with various organizations, and indexing books. Prior to his teaching career he was head of the science and engineering library division at Columbia, which followed serving as librarian in sci-tech units in industry for more than 10 years. He served as editor for 10 years for *Science & Technology Libraries*; since then he has become editor of *Sci-Tech News*, a newsletter sponsored by four divisions of the Special Libraries Association. In 1990 he was added to the SLA Hall of Fame. He has an MS in Physics from Northwestern University, the MLS degree from the University of Illinois, and the DLS degree from Columbia University.

BEATRICE KOVACS

Beatrice Kovacs joined the faculty of the Department of Library and Information Studies, School of Education, University of North Carolina at Greensboro, in 1985 to teach courses in bibliography and literature of science and technology, collection management, organizing library collections, and special libraries. Prior to teaching, she was collection development officer at two academic health science libraries, as well as a cataloger and assistant acquisition librarian at a regional public library. Her experience in library-related work spans more than 20 years. She coauthored, under the last name of Basler, an annotated bibliography called *Health Science Librarianship* for Gale Research in 1977 and in 1990 published a book entitled *The Decision Process for Library Collections: Case Studies in Four Types of Libraries* with Greenwood Press. She received her MLS from Rutgers University and her DLS from Columbia University.